PHYSICAL CLIMATOLOGY

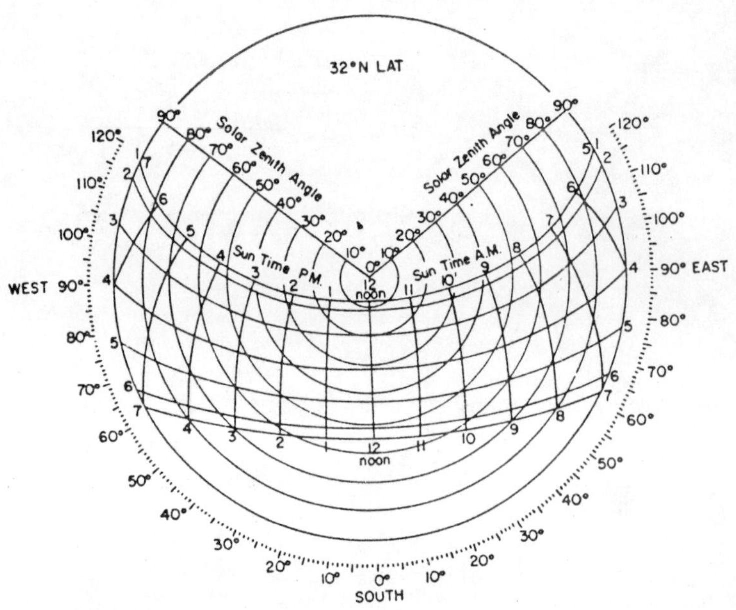
32°N LAT
Solar Zenith Angle
Solar Zenith Angle
Sun Time P.M.
Sun Time A.M.
12 noon
WEST 90°
90° EAST
SOUTH

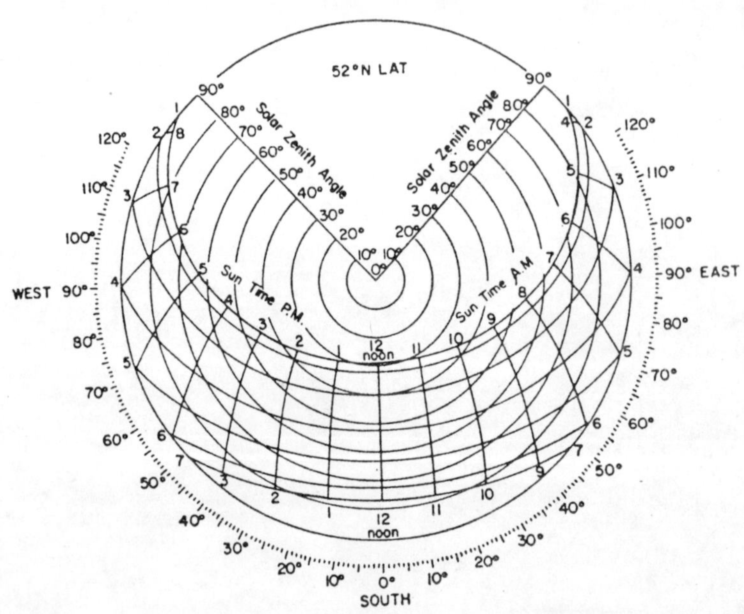
52°N LAT
Solar Zenith Angle
Solar Zenith Angle
Sun Time P.M.
Sun Time A.M.
12 noon
WEST 90°
90° EAST
SOUTH

WILLIAM D. SELLERS

PHYSICAL CLIMATOLOGY

NEW INDIA PUBLISHING AGENCY
New Delhi – 110 034

THE UNIVERSITY OF CHICAGO PRESS, CHICAGO 60637

The University of Chicago Press, Ltd., London

Published 1965. Fourth Impression 1972

With Permission from The University of Chicago Press, Chicago

First Indian Print 2015

ISBN: 978-93-83305-57-5

NEW INDIA PUBLISHING AGENCY

101, Vikas Surya Plaza, CU Block, LSC Market
Pitam Pura, New Delhi 110 034, India
Phone: + 91 (11)27 34 17 17 Fax: + 91(11) 27 34 16 16
Email: info@nipabooks.com
Web: www.nipabooks.com

PREFACE

This text is based on notes prepared for a course in physical climatology taught at The University of Arizona since 1960. Class members come from a wide range of disciplines, including meteorology, hydrology, watershed and range management, agricultural chemistry and soils, agricultural economics, botany, zoology, electrical and civil engineering, geography, and geochronology. Most of the students are in the Graduate College, and all have had at least an introductory course in meteorology. The mathematical preparation of the students varies considerably. Some have carried their training through boundary value problems and complex variables; others have had little more than college algebra and have done poorly in that.

To teach a course that would be useful and interesting to all of these students turned out to be almost impossible. Originally, the English translation of M. I. Budyko's *Teplovoi Balans Zemnoi Poverkhnosti* (The Heat Balance of the Earth's Surface) was used as a text. This was not very satisfactory, however, partly because it is difficult to interpret some of the translation, but mainly because much of the material has to be taken on faith, since it comes from Russian papers that have not been translated into English and are not readily available.

The decision to drop Budyko's monograph and to prepare my own notes was influenced greatly by the publication, in 1961, of the fourth edition of R. Geiger's *Das Klima der bodennahen Luftschicht.* This book, greatly improved since earlier editions but only recently available in English, contains much useful information, some of which was translated and presented to my class.

Those who are familiar with these works by Budyko and Geiger will find their influence time and again in this book. It is only proper, then, that I acknowledge here the indirect assistance of these two climatologists. Dr. Geiger also contributed directly, suggesting the development given in Appendix 3.

For permission to use diagrams, tables, and data from published papers I am grateful to H. H. Lettau (Tables 2 and 3), the Libbey-Owens-Ford Glass Company (Fig. 4), E. C. Kung, R. A. Bryson, and D. H. Lenschow (Table 5), the Controller of H.M. Stationery Office (Figs. 20–23), the North-Holland Publishing Company (the physical data in chap. 9 taken from *Physics of Plant Environment*, edited by W. R. van Wijk), J. Kondo (the physical data for Lake Towada in Table 21), I. C. McIlroy and D. E. Angus (Table 22 and part of Table 23), E. W. Bierly and E. W. Hewson (Fig. 46), and R. W. Fairbridge (Fig. 47).

I would also like to thank Donald D. Perceny for drafting most of the figures, Mrs. Gail Bateman and Mrs. Helen Hindman for typing the manuscript, and Dr. A. R. Kassander, Jr., for his encouragement and technical advice.

William D. Sellers

The University of Arizona

CONTENTS

1 / INTRODUCTION

There has long existed a need for a text devoted primarily to physical climatology, with special emphasis on the global energy and water balance regimes of the earth and its atmosphere. Although these topics are often discussed in textbooks on physical and dynamical meteorology, such as those of J. E. Johnson (1954) and Haltiner and Martin (1957), the treatments are usually very sketchy and incomplete.

The most complete discussion is given by Budyko (1956) in an excellent Russian monograph, reviewed in detail by Malkus (1962). Budyko, however, deals mainly with processes taking place near the earth's surface and presents only a limited discussion of the role of the atmosphere in the global energy balance. Further, much of the data given in the monograph has been revised by Budyko and his colleagues.

Since the water balance and the closely associated hydrologic cycle form the basis for most water conservation practices, their importance and significance are immediately apparent to students. On the other hand, the role of the energy balance is not always so apparent and the question often arises why it is given so much attention. First, virtually all physical processes occurring near the ground or in the atmosphere involve energy transformations or transfer. It would not be exaggerating to say that the physical state of any entity, from a small leaf to a whole continent, is determined by the way it utilizes the available energy. For this reason, energy balance considerations are important in many design problems and in the artificial modification of climates.

Second, the energy balance provides a rational physical basis for development and interpretation of theories of climatic change. Without some concept of the complex energetic interactions that exist between the earth's surface and the atmosphere and between the polar and equatorial latitudes, it would be almost impossible to arrive at a climatic model that will be physically consistent in all respects.

Finally, since evaporation and transpiration require an energy source, the water and energy balances are intimately related. To study only one, at the exclusion of the other, is impossible for a complete understanding of the subject. The interaction between the two is especially apparent in the atmosphere, where vapor flux and latent heat transfer are synonymous. Further, a significant portion of the warm air that periodically advances poleward in middle latitudes during the winter is derived from energy released by condensation in the tropics.

A thorough study of energy and water balance relationships also entails some discussion of radiation, heat transfer in soil, water, and air, and evapotranspiration. Except for the first, these are more logically topics in micrometeorology or microclimatology than in physical climatology. They also involve considerable mathematics. Nevertheless, because of their importance, they are included here. Many meteorologists are not familiar with the relatively simple principles of heat transfer in soil and water, while nonmeteorologists are likely to have only a vague understanding of convective heat transfer in air.

Because of the current interest in air pollution, a chapter on atmospheric diffusion is also included. This subject is a good example of a practical application of the basic principles of atmospheric turbulence. It also illustrates the fundamental difficulties a meteorologist has in trying to construct a theoretical model of an atmospheric process. So far, diffusion theory has relied heavily on mean-value statistics and has been incapable of yielding much more than idealized probability distributions of pollution dispersal. Perhaps the biggest contribution meteorologists have made to the pollution problem is to present a clear delineation of those atmospheric conditions most conducive to the rapid dissemination of gases, fumes, and other effluents.

This text concludes with a chapter on paleoclimatology and theories of climatic change. The theories should be read with the discussions of the radiation and energy balances (chaps. 5 and 8) firmly in mind. Within any geological age the entire earth-atmosphere system is very nearly in both thermal and radiative equilibrium and its mean temperature does not change appreciably from one year to the next. Any postulated latitudinal distribution of temperature or radiation imbalance must satisfy this condition. Hence, for example, an assumed increase in the output of energy from the sun must be accompanied either by an increase in the planetary albedo or by an increase in the mean temperature of the earth. In either case, the absorbed solar energy must be exactly balanced by the emitted infrared energy. Since an increased albedo, owing to increased cloud and snow cover, implies a cold climate and increased temperatures imply a warm climate, it follows that the same cause could produce two diametrically opposed effects. This, of course, is one reason why the entire field of climatic change is so controversial and speculative.

Except in a very general way, little attention is given here either to the regional distribution of climates or to what might properly be called bioclimatology. The former is covered adequately by Haurwitz and Austin (1944), Kendrew (1953), and Trewartha (1954, 1961), and others, and the latter by Landsberg (1958). It is assumed that most readers are already acquainted with these books and are aware of the regional climatic peculiarities that are often encountered. The relatively simple mean annual latitudinal distributions shown in Figures 7, 14, 19, 26, 29, 34, and 35 often represent the summation or average of these peculiarities.

The order in which the material is presented is the one found most satisfactory for a text. The apparently discontinuous insertion of the discussion of the water balance and the hydrologic cycle between the discussions of radiation instruments and the energy balance was done because students seem to have a much easier time understanding the energy balance if the water balance is discussed first. There may be some merit in covering paleoclimatology and theories of climatic change before atmospheric diffusion, since the latter is a rather specialized subject and requires considerably more mathematics and statistics than most of the rest of the material presented.

Although calculus is used, mainly in chapters 9, 10, and 12, emphasis is placed on the physical rather than the mathematical implications. Several derivations not vital to an understanding of the subject, but still useful as background material, are presented in the appendixes. These should not be covered in class unless time permits.

2 / THE CLIMATOLOGY OF THE HEMISPHERES

This chapter summarizes some of the better known climatological features of the two hemispheres. The last seven columns of Table 1 give the mean annual surface temperature (T, degrees Kelvin), precipitation (r, millimeters), evaporation (E, millimeters), precipitable water vapor (w_a, millimeters), precipitation efficiency (PE, percent), cloud cover (n, percent), and shortwave albedo (a, percent) for each ten-degree latitude zone, for each hemisphere, and for the globe as a whole. Many of the differences noted between the two hemispheres within the same latitude zone may be traced either to differences in the percentage of the zone covered by oceans (O, percent) or to the mean elevation of the zone (z, meters), listed in the first two columns of the table.

Oceans cover 81 percent of the southern hemisphere but only 61 percent of the northern hemisphere. The continents of the northern hemisphere lie mainly between 40 and 70°N and the highest elevations between 20 and 50°N (the Himalayas, Alps, and Rockies). In the southern hemisphere the continents are located north of 40°S and south of 60°S. Except for the Andes of South America, the highest elevations, exceeding 3,000 m, are found on Antarctica.

Temperatures given in Table 1 are estimated mean annual surface temperatures and are lower than the sea level values often quoted in the meteorological literature (Haurwitz and Austin, 1944; Trewartha, 1954), especially between 20 and 50°N and south of 60°S. The average global surface temperature is about 286°K, compared with 288°K for the average global sea level temperature.

When the two hemispheres are compared, it can be seen that temperatures are lower in the northern hemisphere than in the southern hemisphere only between 40 and 60°, where the controlling factor seems to be the high elevations of the northern hemisphere coupled with the frequent intrusion of arctic air into the region during the winter. The relatively high surface temperatures of the northern hemisphere in the tropics and subtropics and especially between 10 and 30°N are directly related to the presence of most of the world's deserts within this zone.

Poleward of 60°S lies the Antarctic continent, which rises to heights of more than 3,570 m. Some idea of the extreme cold of this huge icecap can be obtained by referring to the temperature record for Vostok, a Russian meteorological station located on a gentle slope of the high central plateau at an elevation of 3,488 m. The

TABLE 1

PHYSICAL AND CLIMATOLOGICAL DATA

Latitude Zone	O (percent)	z (m)	T (° K)	r (mm)	E (mm)	w_a (mm)	PE (percent)	n (percent)	a (percent)
80–90° N. .	93.4	137	249.6	120	42	4.90	6.7	62	61
70–80.	71.3	220	257.3	185	145	6.48	7.8	66	46
60–70.	29.4	202	266.0	415	333	8.52	13.3	65	24
50–60.	42.8	296	273.7	789	469	11.64	18.6	60	14
40–50.	47.5	382	280.7	907	641	15.21	16.3	53	12
30–40.	57.2	496	287.2	872	1,002	18.95	12.6	46	10
20–30.	62.4	366	293.6	790	1,246	26.37	8.2	43	10
10–20.	73.6	146	298.3	1,151	1,389	36.73	8.6	47	9
0–10.	77.2	158	298.7	1,934	1,235	41.07	12.9	52	8
0–90° N.	60.6	284	286.4	1,009	944	23.85	12.1	52	14
0–10° S. . .	76.4	154	298.0	1,445	1,304	40.90	9.7	52	7
10–20	78.0	121	296.5	1,132	1,541	36.66	8.5	48	8
20–30.	76.9	156	292.0	857	1,416	29.86	7.9	48	8
30–40.	88.8	106	286.7	932	1,256	23.81	10.7	54	8
40–50.	97.0	5	281.9	1,226	895	18.10	18.6	66	7
50–60.	99.2	5	274.4	1,046	520	12.61	22.7	72	17
60–70.	89.6	388	262.2	418	174	6.84	16.7	76	21
70–80.	24.6	1,420	243.7	82	45	2.87	7.8	65	63
80–90.	0.0	2,272	225.3	30	0	1.56	5.3	54	84
0–90° S. .	80.9	216	284.6	1,000	1,064	25.49	12.1	57	13
Globe.	70.8	250	285.5	1,004	1,004	24.67	12.1	54	13

average temperature of the warmest month (January) is −32.5°C and of the coldest month (August), −69.8°C. The lowest temperature ever observed on the earth's surface, −88.3°C, was recorded here on August 24, 1960 (Stepanova, 1963). Even on the warmest summer days, the temperature rarely rises above −25°C. The cold is accentuated by the strong winds that prevail on the continent. At Vostok the average speed is about 5 m sec^{-1}, with extremes up to 25 m sec^{-1}. Mirny, at an elevation of 35 m on the edge of the continental ice sheet on the Antarctic Circle, has a milder climate, the average annual temperature being −11°C, but the winds are very strong, averaging 12 m sec^{-1} and ranging up to 50 m sec^{-1} (Cartwright and Rubin, 1961).

Both hemispheres receive roughly the same average annual precipitation (1,000 mm). The slightly greater amount in the northern hemisphere is produced within the intertropical convergence zone, which has a mean location slightly north of the equator (Trewartha, 1954). Actually, according to these figures, which have been derived from data presented by Brooks and Hunt (1930), Meinardus (1934), Wüst (1954), Budyko (1956), and Budyko *et al.* (1962), the northern hemisphere is drier than the southern hemisphere between 20 and 70°.

Precipitation in both hemispheres is highest near the equator and between 40 and 50°, the latitudes of frequent cyclonic storm activity. It is lowest near the poles and in the subtropics between 20 and 30°. Of course, there are significant longitudinal variations from this average pattern. For example, parts of Italian Somaliland on the equator in East Africa have an annual rainfall of less than 200 mm, while the world's greatest average annual precipitation, 12,000 mm, occurs on Mt. Waialeale, Hawaii, at 22°5′N in the subtropics.

Evaporation rates are highest in the dry subtropics and decrease toward both the equator and the poles. The peak is much more pronounced in the southern hemisphere than in the northern hemisphere, whose subtropical desert land areas are incapable of adding much moisture to the atmosphere. The significantly greater evaporation from the southern hemisphere as a whole is directly related to the greater ocean area of that hemisphere. The difference is not as great as might be expected because the relatively low temperatures and considerable cloud cover of the southern hemisphere tend to suppress evaporation. No significant evaporation can occur from the Antarctic and Greenland icecaps because of the extremely low temperatures that prevail there.

Precipitation exceeds evaporation poleward of 40° in both hemispheres and in the tropics between 10°N and 10°S. Since no latitude belt seems to be getting wetter or drier, this implies that moisture must be transferred to these regions from the subtropical latitude zones where evaporation exceeds precipitation. In addition, there must be a net transfer of moisture from the southern hemisphere, where evaporation exceeds precipitation, to the northern hemisphere, where precipitation exceeds evaporation. A significant portion of this transfer may occur in the Indian Ocean during the summer monsoon season.

The precipitable water vapor may be defined as the amount of water vapor, usually expressed in centimeters or millimeters, in a vertical column of air with a cross-section of 1 cm^2 and a fixed depth. In Table 1 average values are given for a column extending from the ground surface to the top of the atmosphere. They have been estimated from maps prepared by Bannon and Steele (1960).

As shown in the table, the precipitable water vapor in the atmosphere decreases from more than 40 mm in the tropics to less than 5 mm near the poles. This decrease is the result of a corresponding poleward decrease of temperature, since the colder the air the more likely the vapor present will condense.

The small amount of water vapor in the atmosphere at high latitudes is accompanied by low precipitation rates. Low precipitation rates over many of the world's deserts, however, are accompanied by very high atmospheric vapor contents. Hence, the distribution of precipitation cannot be explained solely by the distribution of water vapor. The methods by which precipitation can be released must also be considered. The atmosphere over deserts is typically very stable; ascending motions and the resultant cooling and condensation are suppressed. On the other hand, the frequent cyclonic activity of the middle latitudes, accompanied by large-scale rising motion and instability, provides excellent conditions for the release of large amounts of moisture.

An idea of the latitudinal variation of the precipitation efficiency can be obtained by dividing the mean daily precipitation ($r/365$) by the average precipitable water vapor. For convenience, this ratio can be thought of as the fraction of the average moisture overhead which falls as precipitation on an average day. The precipitation efficiency is greatest in the high middle latitudes of both hemispheres (Table 1), where cyclonic storms are most active, and between the equator and 10°N, within the intertropical convergence zone. It is least near the poles and in the subtropics, where the stability of the air is pronounced. For the globe as a whole, roughly 12 percent of the moisture overhead falls on an average day, giving a turnover time for atmospheric water vapor of about 8 days.

In the United States in January (Fig. 1) the precipitation efficiency is low on the leeward side of the major mountain ranges and high on the windward side of the same ranges, reflecting the orographic effect on precipitation. Highest values, 50 to 80 percent, occur along the coasts of Maine and Washington. Lowest values, 4 to 5 percent, are found in the Southwest. Precipitation efficiencies are generally lower in July (Fig. 2) than in January, with the lowest values along the Pacific Coast, which is under the influence of the very stable Pacific high-pressure system of summer. Values in excess of 10 percent are found only in the southeastern states and along the northeastern border.

The cloud cover values given in Table 1 are those obtained 40 years ago by Brooks (1927), modified by more recent data given for the northern hemisphere by Houghton (1954). Except in the polar regions, cloudiness is greater in the southern hemisphere than in the northern hemisphere. This is related to the relatively low temperatures, high evaporation rates, and excessive air moisture content of the southern hemisphere.

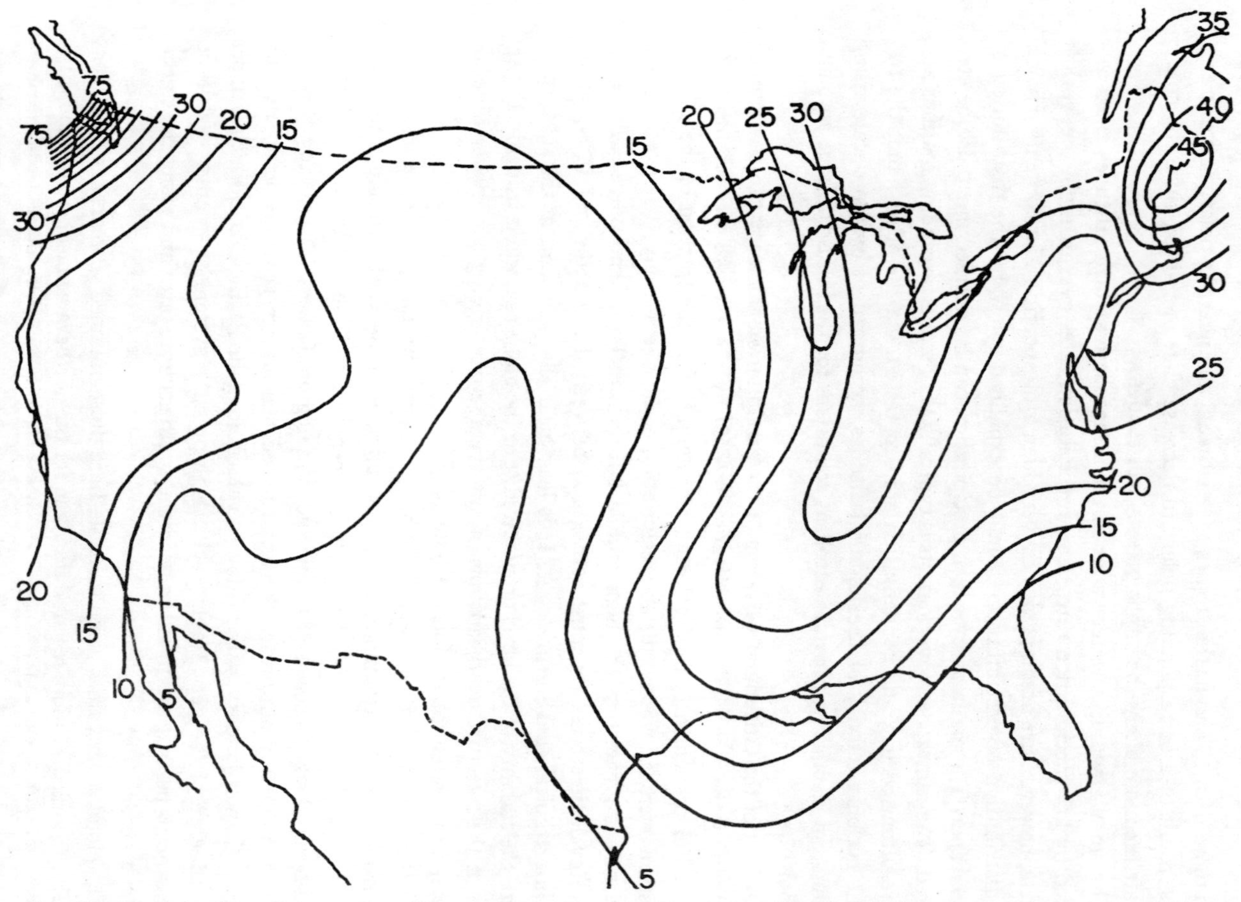

Fig. 1.—Precipitation efficiency (percent) in the United States in January. The precipitable water vapor values used to construct this figure were taken from Reitan (1960*a*).

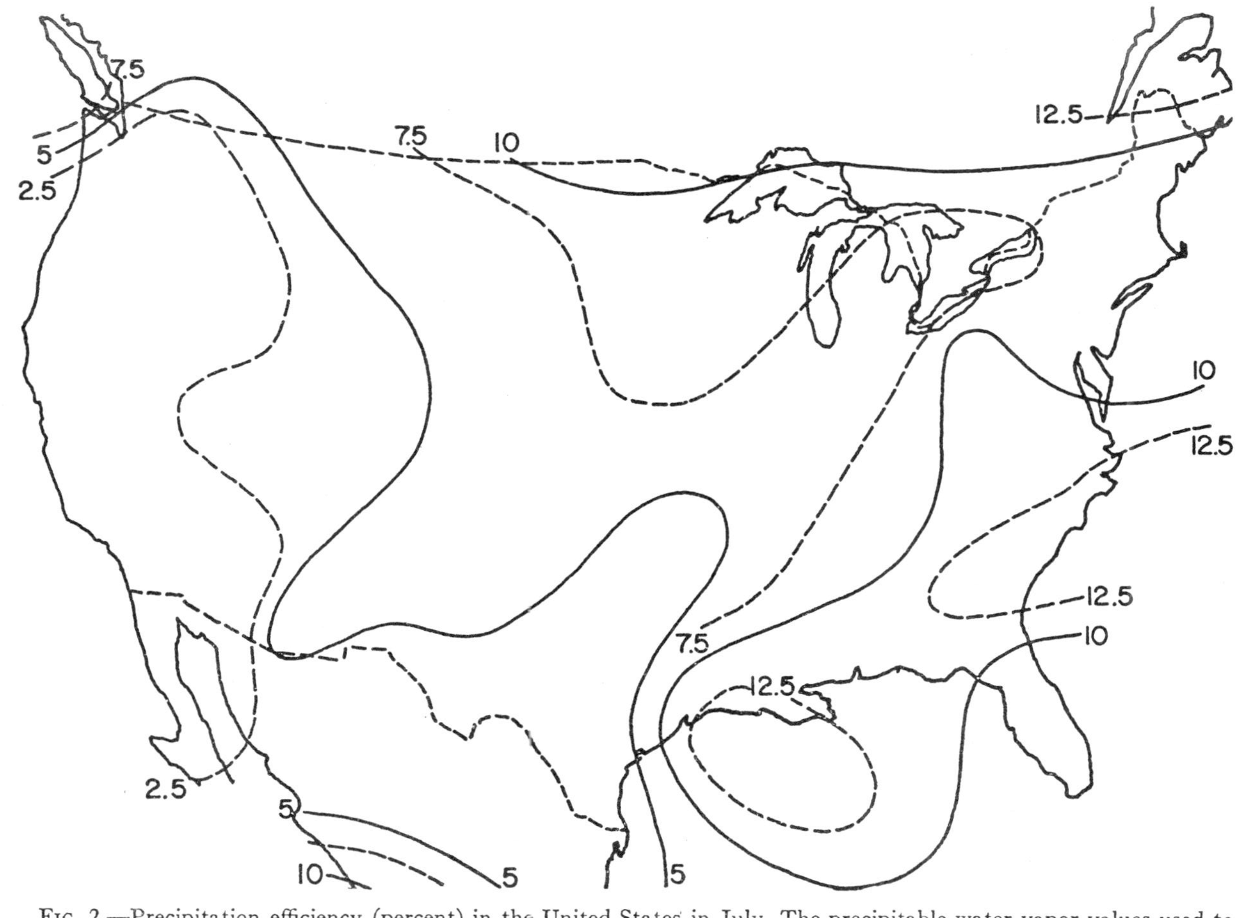

FIG. 2.—Precipitation efficiency (percent) in the United States in July. The precipitable water vapor values used to construct this figure were taken from Reitan (1960*a*).

The average annual surface albedo for each latitude zone is given in the last column of Table 1. This quantity is defined as the percentage of the incident shortwave (solar) radiation that is reflected from the ground. The data were obtained from Gabites (1950), Houghton (1954), and Hanson (1960). A detailed discussion of the surface albedo will be given in the following chapter. For the time being, it is important to note that the surface albedo is principally a function of the color of the absorbing surface and angle of incidence of the solar radiation. Latitudinally it varies greatly, from 84 percent on the Antarctic icecap to 7 to 8 percent between 10°N and 50°S, a region consisting largely of oceans and tropical rain forests.

3 / SOLAR RADIATION

In the remainder of this book the metric system of units will be used. The basic unit of energy will be one 15°C gram-calorie (cal), which is the amount of heat required to raise the temperature of 1 g of water 1°C from 14.5°C to 15.5°C. Energy per unit area will be expressed in langleys or kilolangleys. One langley (ly) is equivalent to 1 cal cm^{-2}. A kilolangley (kly) is equal to 1,000 ly. The conversion to other energy units may be accomplished with the aid of Appendix 1.

The sun provides about 99.97 percent of the heat energy required for the physical processes taking place in the earth-atmosphere system. Each minute it radiates approximately 56×10^{26} cal of energy. In terms of the energy per unit area incident on a spherical shell with a radius of 1.5×10^{13} cm (the mean distance of the earth from the sun) and concentric with the sun, this energy is equal to

$$S = \frac{56 \times 10^{26} \text{ cal min}^{-1}}{4\pi(1.5 \times 10^{13} \text{ cm})^2} = 2.0 \text{ ly min}^{-1}$$

where S is the solar constant.

The solar constant is normally defined as the flux of solar radiation at the outer boundary of the earth's atmosphere that is received on a surface held perpendicular to the sun's direction at the mean distance between the sun and earth. From the above equation, however, it may also be defined as the flux of solar radiation on a spherical shell concentric with the sun whose radius is equal to the mean distance between the sun and earth. In both cases the term "flux" refers to the total radiant energy of all wavelengths crossing a unit area of surface per unit time.

The above value for the solar constant, which will be used throughout this text, is that obtained by F. S. Johnson (1954) and is slightly higher than the value of 1.94 ly min^{-1} often quoted in the literature. The solar constant is not a true constant but fluctuates by as much as ± 1.5 percent about its mean value. Much of the fluctuation seems to be in the ultraviolet portion of the solar spectrum (Rense, 1961).

The total solar energy intercepted by the earth in unit time is equal to $\pi a^2 S$, where a is the radius of the earth (6.37×10^8 cm). Hence,

$$\pi a^2 S = 2.55 \times 10^{18} \text{ cal min}^{-1} = 3.67 \times 10^{21} \text{ cal day}^{-1}.$$

If this energy is spread uniformly over the full surface of the earth, the amount received per unit area and time at the top of the atmosphere is

$$\bar{Q}_s = \frac{\pi a^2 S}{4\pi a^2} = \frac{S}{4} = 0.5 \text{ ly min}^{-1} = 263 \text{ kly year}^{-1}.$$

Actually the distribution is not uniform; the annual value at the equator is about 2.4 times that near the poles. The above value, however, is a realistic average for the globe as a whole.

TABLE 2

LARGE-SCALE ENERGY SOURCES

[Rates are relative to solar energy available (263 kly year^{-1}).]

Source	Rate
One-fourth solar constant	1
Heat flux from the earth's interior	18×10^{-5}
Infrared radiation from the full moon	3×10^{-5}
Sun's radiation reflected from the full moon	1×10^{-5}
Energy generated by solar tidal forces in the atmosphere	1×10^{-5}
Combustion of coal, oil, and gas in the United States	7×10^{-6}
Energy dissipated in lightning discharges	6×10^{-7}
Dissipation of magnetic storm energy	2×10^{-7}
Radiation from bright aurora	14×10^{-8}
Energy of cosmic radiation	9×10^{-8}
Dissipation of mechanical energy of micrometeorites	6×10^{-8}
Total radiation from stars	4×10^{-8}
Energy generated by lunar tidal forces in the atmosphere	3×10^{-8}
Radiation from zodiacal light	1×10^{-8}

Most data in this table and in Table 3 have been obtained from an unpublished series of notes by H. H. Lettau (Department of Meteorology, University of Wisconsin).

In Table 2 the energy from the sun intercepted by the earth is compared with other large-scale energy sources that act continuously or quasicontinuously in the atmosphere and at its boundaries. Rates are given relative to the solar energy available. Note the overpowering dominance of solar radiation.

In Table 3 some estimates are made of the total energy involved in various individual phenomena or localized processes in the atmosphere. Rates are relative to the energy received by the earth from the sun in one day (3.67×10^{21} cal). This energy is sufficient to generate 10 thousand hurricanes, 100 million thunderstorms, or 100 billion tornadoes. If it were all collected and stored, there would be enough energy available to satisfy the world's industrial and domestic needs for 100 years. Of course, one would not want to do this, even if it were possible, because no energy would be left to warm the air and evaporate water, and both the general circulation of the atmosphere and the hydrologic cycle would come to a halt.

TABLE 3

TOTAL ENERGY OF VARIOUS INDIVIDUAL PHENOMENA AND LOCALIZED PROCESSES IN THE ATMOSPHERE

[Rates are relative to total solar energy intercepted by the earth (3.67×10^{21} cal day^{-1}).]

Solar energy received per day	1
Melting of average winter snow during the spring season	10^{-1}
Monsoon circulation	10^{-2}
World use of energy in 1950	10^{-2}
Strong earthquake	10^{-2}
Average cyclone	10^{-3}
Average hurricane	10^{-4}
Krakatoa explosion of August, 1883	10^{-5}
Detonation of "thermonuclear weapon" in April, 1954	10^{-5}
Kinetic energy of the general circulation	10^{-5}
Average squall line	10^{-6}
Average magnetic storm	10^{-7}
Average summer thunderstorm	10^{-8}
Detonation of Nagasaki bomb in August, 1945	10^{-8}
Average earthquake	10^{-8}
Burning of 7,000 tons of coal	10^{-8}
Daily output of Hoover Dam	10^{-8}
Moderate rain (10 mm over Washington, D.C.)	10^{-8}
Average forest fire in the United States, 1952–53	10^{-9}
Average local shower	10^{-10}
Average tornado	10^{-11}
Street lighting on average night in New York City	10^{-11}
Average lightning stroke	10^{-13}
Average dust devil	10^{-15}
Individual gust near the earth's surface	10^{-17}
Meteorite	10^{-18}

SOLAR RADIATION AT THE TOP OF THE ATMOSPHERE

The amount of solar radiation actually incident on the top of the atmosphere depends on the time of year, the time of day, and the latitude. If A_h is an element of area parallel to the earth's surface at the top of the atmosphere and A_n is the projection of this area on a plane normal to the sun's rays at any instant, it follows that $\cos Z = A_n/A_h$ where Z is the zenith angle of the sun, that is, the angular distance of the sun from the local vertical. Because the areas are at the top of the atmosphere where there is no absorbing medium, the amount of energy that passes through the surface A_h

must equal what passes through A_n. Hence, from the definition of the solar constant, $S(\bar{d}/d)^2 A_n = Q_s' A_h$ where d and $\bar{d}$ are, respectively, the instantaneous and mean distances of the earth from the sun and Q_s' is the instantaneous flux of solar radiation through the area A_h. Combining the above two equations,

$$Q_s' = S\left(\frac{\bar{d}}{d}\right)^2 \cos Z\,. \qquad (3.1)$$

The zenith angle of the sun is usually not measured directly and must be determined from other angles that are known. In Figure 3, which is derived from Humphreys (1940), let P be the point of observation at the top of the atmosphere and OV the zenith through this point. Then, if the sun is in the direction OS or PS, the plane of OV and OS will intersect the surface of the earth in a great circle and the angle VOS, measured by the arc PX of this circle, is equal to the sun's zenith distance Z. In the spherical triangle NPX, the arc NX (X being directly under the sun) is equal

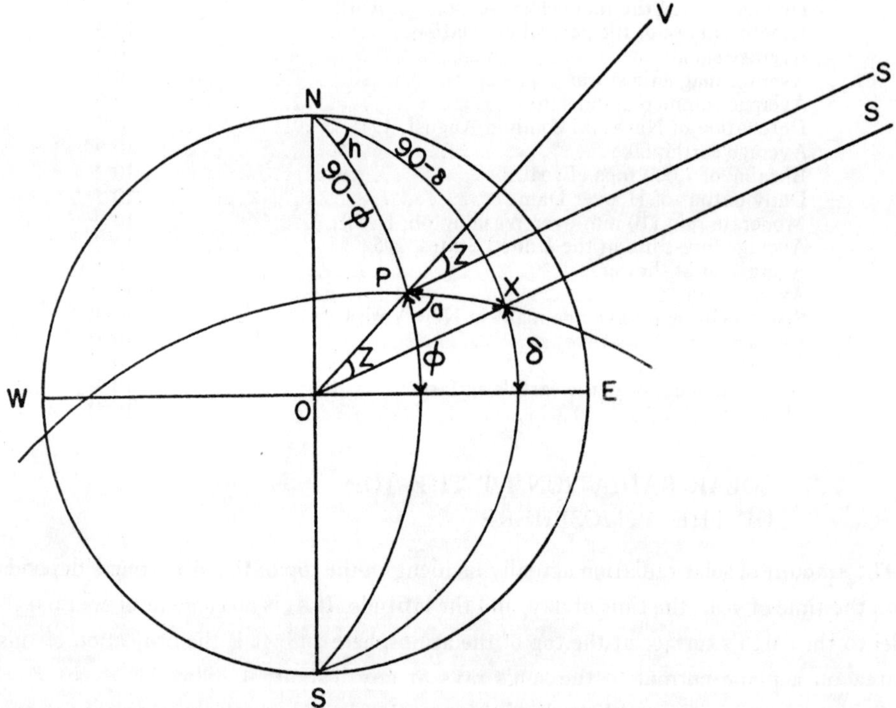

FIG. 3.—Relation of the solar zenith angle Z to the latitude φ, the solar declination δ, the hour angle h, and the azimuth angle a of the sun from south.

to 90° minus the solar declination δ, which is the angular distance of the sun north (positive) or south (negative) of the equator. Finally, the arc NP is equal to 90° minus the latitude φ of the observation point and the angle h is the hour angle or the angle through which the earth must turn to bring the meridian of P directly under the sun. ϕ is negative in the southern hemisphere and h is negative in the morning. From spherical trigonometry it then follows that

$$\cos Z = \sin \varphi \sin \delta + \cos \varphi \cos \delta \cos h \,. \tag{3.2}$$

The solar declination is a function only of the day of the year and is independent of the location of the observation point. It varies from 23°27′ on June 21 to −23°27′ on December 22. Values for each day and hour of the year may be obtained from *The Nautical Almanac*, published annually by the U.S. Government Printing Office in Washington, D.C. The hour angle is zero at solar noon, that is, when the sun is directly north or south of the observation point, and increases by 15° for every hour before or after solar noon. At 0600, for example, the hour angle is −90°.

Several interesting results may be derived from equation (3.2).

1. At the poles $\cos \varphi = 0$, $\sin \varphi = 1$, and $\cos Z = \sin \delta$ or $90° - Z = \delta$. Hence, at these points the elevation angle of the sun always equals the declination angle and during the 6 months of daylight the sun simply circles around the horizon, never rising more than 23.5°. The transition from day to night occurs at the equinoxes (March 20 and September 23) when the declination is 0° and the path of the sun coincides with the horizon.
2. At solar noon at any latitude $\cos h = 1$ and $Z = \varphi - \delta$.
3. At sunrise or sunset at any latitude except the poles $\cos Z = 0$, $h = H =$ half-day length, and

$$\cos H = -\tan \varphi \tan \delta \,. \tag{3.3}$$

The half-day length will be 6 hours if either $\tan \varphi = 0$ (the equator on all days) or $\tan \delta = 0$ (the equinoxes at all latitudes except the poles). The latitude of the polar night may be found by setting $H = 0$ in equation (3.3). Then $\tan \varphi = -\cot \delta$ $(\delta \neq 0)$ and the latitude of the polar night is equal to $90° - |\delta|$ in the winter hemisphere.

In Figure 3, the azimuth angle of the sun from the south, a, can be obtained by applying the law of sines to the spherical triangle NPX, with the result that

$$\sin a = \frac{\cos \delta \sin h}{\sin Z} \tag{3.4}$$

or alternately from the law of cosines

$$\cos a = \frac{\sin \varphi \cos Z - \sin \delta}{\cos \varphi \sin Z}. \tag{3.5}$$

At sunrise or sunset $\cos Z = 0$, $\sin Z = 1$, $h = H$, and

$$\sin a_0 = \cos \delta \sin H \quad \text{or} \quad \cos a_0 = -\frac{\sin \delta}{\cos \varphi}. \tag{3.6}$$

For any given solar declination, the azimuth angle of the sun at sunrise will be most nearly due east at the equator and depart more and more therefrom with increasing latitude. During the northern hemisphere winter, the sun will rise and set almost due south just equatorward of the latitude $90° + \delta$ in both hemispheres. During the northern hemisphere summer, it will rise and set almost due north just equatorward of the latitude $90° - \delta$ in both hemispheres. At the equator the azimuth angle of the sun at sunrise is always equal to $90° + \delta$. These conclusions are the result of the tilt of the earth's axis relative to its plane of rotation around the sun.

A convenient way to visualize the results of the preceding paragraphs is to refer to a series of sun-path diagrams, such as those presented in the *Smithsonian Meteorological Tables* (List, 1958). Two of these, for latitudes 32°N and 52°N, are reproduced in Figure 4. To find the zenith angle and azimuth of the sun at 11 A.M. solar time on February 21 at 32°N, path line Number 5 in the upper part of the figure is followed to its intersection with the 11 A.M. line. At this point the zenith angle is 45° and the azimuth 21° east of south.

The daily total solar radiation incident on a horizontal surface at the top of the atmosphere Q_s can be determined by summing or integrating equation (3.1) from sunrise to sunset. The result, derived in Appendix 2, is

$$Q_s = \frac{1{,}440}{\pi} S \left(\frac{\bar{d}}{d}\right)^2 (H \sin \varphi \sin \delta + \cos \varphi \cos \delta \sin H) \text{ly day}^{-1} \tag{3.7}$$

where H in the first term on the right is expressed in radians ($90° = 0.5\pi$ rad). The factor $(\bar{d}/d)^2$ never departs by more than 3.5 percent from unity, ranging from 1.0344 on January 3 to 0.9674 on July 5.

The daily variation of Q_s as a function of latitude is shown in Figure 5. Because the sun is closest to the earth in January, the distribution is slightly asymmetric and more radiation is received at all latitudes during the northern hemisphere winter than during the northern hemisphere summer. The peak value, 1,185 ly day^{-1}, occurs at the South Pole on December 22.

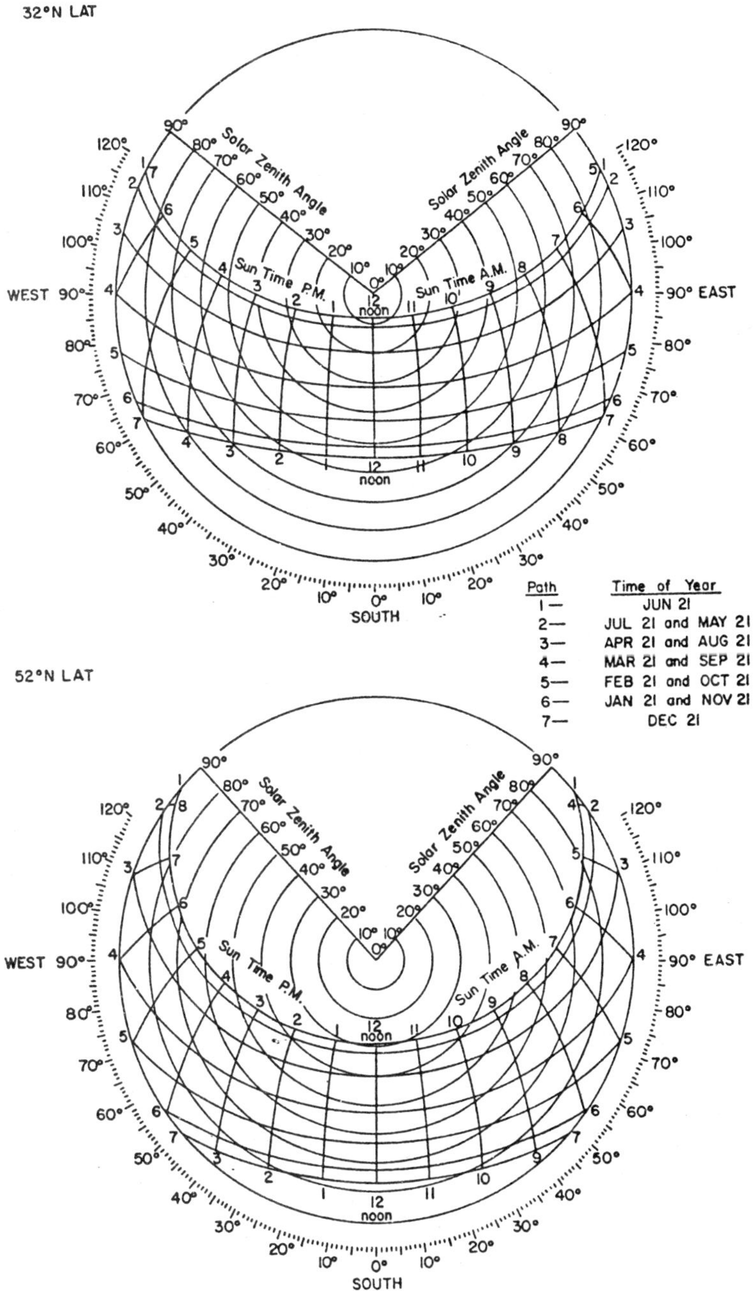

FIG. 4.—Sun-path diagrams for 32°N and 52°N. (Reproduced with the permission of the Libbey-Owens-Ford Glass Company.)

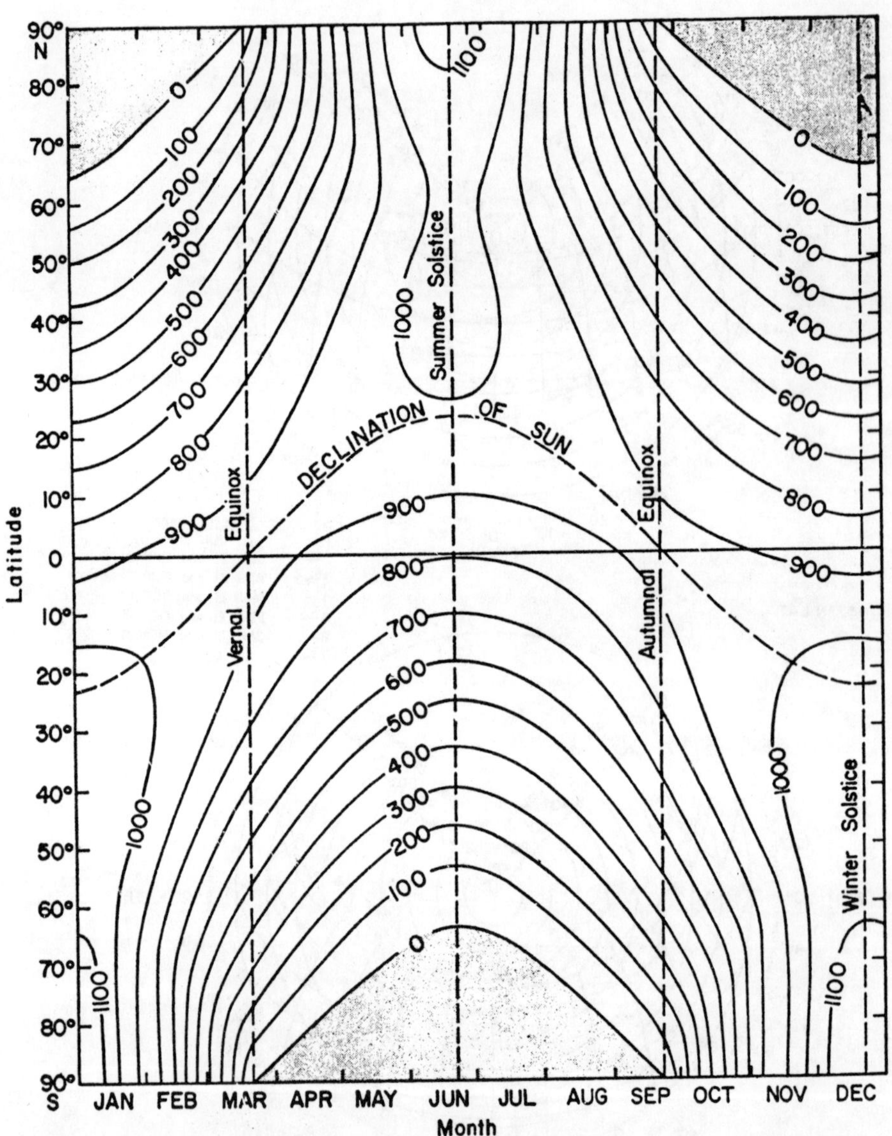

FIG. 5.—The daily variation of the solar radiation at the top of the atmosphere as a function of latitude. The units are langleys per day. Modified from List (1958).

NATURE OF SOLAR RADIATION

If the sun is assumed to be a black body, that is, one which absorbs all the radiation falling upon its surface, an estimate of its effective radiating temperature may be obtained from the Stefan-Boltzmann law. This law states that the flux of radiation from a black body is directly proportional to the fourth power of its absolute temperature; that is,

$$\text{Flux} = \sigma T^4 \qquad (3.8)$$

where σ is the Stefan-Boltzmann constant (8.14×10^{-11} ly min^{-1} $K^{-4} = 1.17 \times 10^{-7}$ ly day^{-1} K^{-4}). In the case of the sun, the flux is equal to 56×10^{26} cal min^{-1} divided by the surface area of the sun (6.093×10^{22} cm^2), or to 9.2×10^4 ly min^{-1}. This gives an effective temperature of about 5,800°K. Actually, as shown in Figure 6, the sun radiates as a 6,000°K black body in the spectral range from 1.2μ to at least 10μ. ($1\mu = 1$ micron $= 10^{-4}$ cm.) At shorter wavelengths, at least to 0.22μ, absorption within the solar atmosphere reduces the radiating temperature to between 4,000 and 5,000°K.

By the Wien displacement law, the wavelength of maximum intensity of emission λ_{max} from a black body is inversely proportional to the absolute temperature T of the body. Thus,

$$\lambda_{max}\ (\mu) = 2{,}897\ T^{-1}\,. \qquad (3.9)$$

For the sun the wavelength of maximum emission is near 0.5μ, which is in the visible portion of the electromagnetic spectrum.

Almost 99 percent of the sun's radiation is contained in the so-called short wavelengths from 0.15 to 4.0μ. Of this, 9 percent is in the ultraviolet ($\lambda \leq 0.4\mu$), 45 percent is in the visible ($0.4\mu \leq \lambda \leq 0.74\mu$), and 46 percent is in the infrared ($\lambda \geq 0.74\mu$).

DISPOSITION OF SOLAR RADIATION IN THE EARTH-ATMOSPHERE SYSTEM

The solar radiation intercepted by the earth will either be absorbed and used in energy-driven processes or be returned to space by scattering or reflection. In mathematical form, the disposition of solar radiation is given by

$$Q_s = C_r + A_r + C_a + A_a + (Q + q)(1 - a) + (Q + q)a\,. \qquad (3.10)$$

This expression says that the solar radiation incident on a horizontal surface at the top of the atmosphere can be reflected and scattered back to space by clouds (C_r), by dry air molecules, dust, and water vapor (A_r), or by the earth's surface ($(Q + q)a$), where Q and q are, respectively, the direct beam and diffuse solar radiation incident

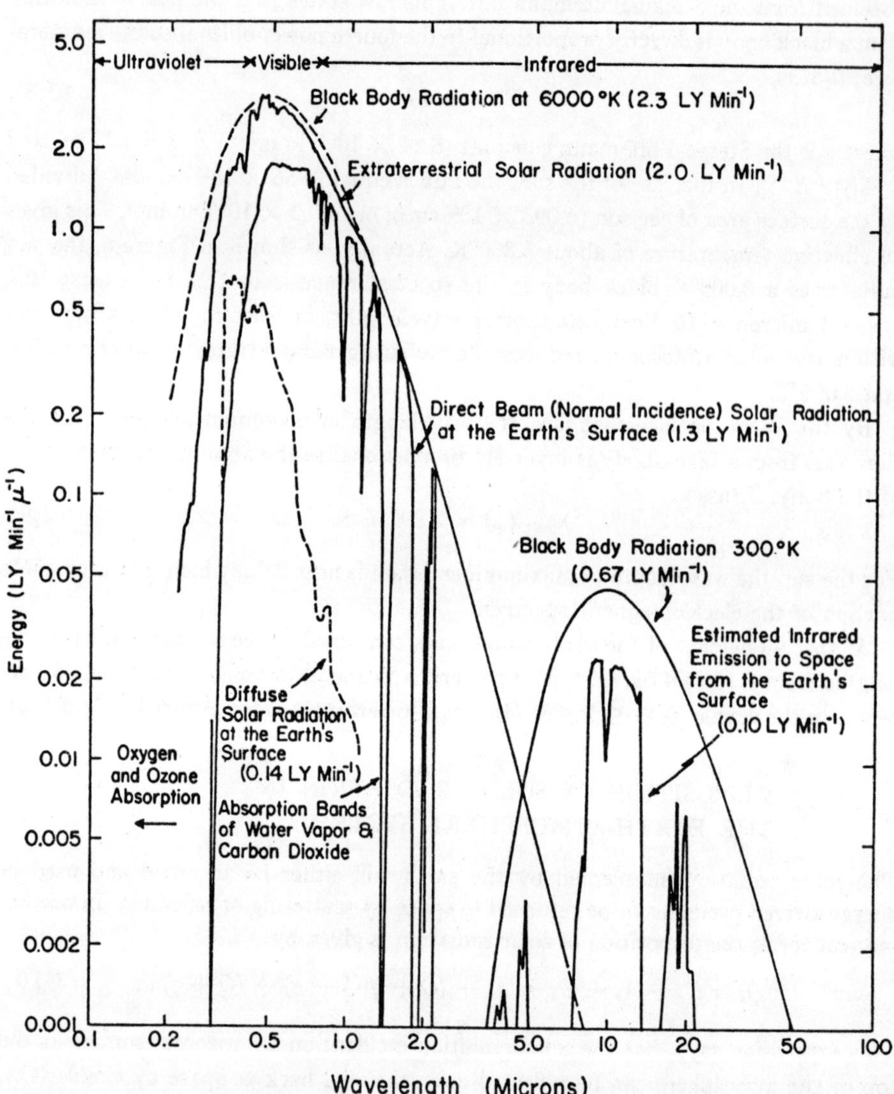

FIG. 6.—Electromagnetic spectra of solar and terrestrial radiation. The black body radiation at 6,000° K is reduced by the square of the ratio of the sun's radius to the average distance between the sun and the earth in order to give the flux that would be incident on the top of the atmosphere.

on a horizontal area at the ground and a is the surface albedo. Alternatively this solar radiation can be absorbed by clouds (C_a), by dry air molecules, dust, and water vapor (A_a), or by the earth's surface $[(Q + q)(1 - a)]$. The average annual latitudinal distribution of each component is shown in the lower part of Figure 7. The distributions of $Q + q$, Q, and q are given in the upper part of the figure. The curves are based mainly on data given by Gabites (1950), Houghton (1954), Budyko (1956), and Hanson (1960), and are for normal cloud cover.

TABLE 4

Albedos for the Shortwave Portion of the Electromagnetic Spectrum

(Wavelengths < 4.0μ)

A. Water Surfaces	
Winter— 0° latitude	6
30° latitude	9
60° latitude	21
Summer— 0° latitude	6
30° latitude	6
60° latitude	7
B. Bare Areas and Soils	
Snow, fresh fallen	75–95
Snow, several days old	40–70
Ice, sea	30–40
Sand dune, dry	35–45
Sand dune, wet	20–30
Soil, dark	5–15
Soil, moist gray	10–20
Soil, dry clay or gray	20–35
Soil, dry light sand	25–45
Concrete, dry	17–27
Road, black top	5–10
C. Natural Surfaces	
Desert	25–30
Savanna, dry season	25–30
Savanna, wet season	15–20
Chaparral	15–20
Meadows, green	10–20
Forest, deciduous	10–20
C. Natural Surfaces (cont.)	
Forest, coniferous	5–15
Tundra	15–20
Crops	15–25
D. Cloud Overcast	
Cumuliform	70–90
Stratus (500–1,000′ thick)	59–84
Altostratus	39–59
Cirrostratus	44–50
E. Planets	
Earth	34–42
Moon	6.7
Jupiter	73
Mars	16
Mercury	5.6
Neptune	84
Pluto	14
Saturn	76
Uranus	93
Venus	76
F. Human Skin	
Blond	43–45
Brunette	35
Dark	16–22

A global average of about one-fourth of the incident radiation is reflected to space by clouds, which have an albedo that is generally greater than 50 percent and may exceed 90 percent for the towering cumulonimbus clouds of summer. See Table 4. The high values of this component in the southern hemisphere are associated with the greater cloudiness of that hemisphere. On a percentage basis, reflection by clouds is greatest in the middle and upper middle latitudes of both hemispheres and least in the subtropics of both hemispheres and near the South Pole.

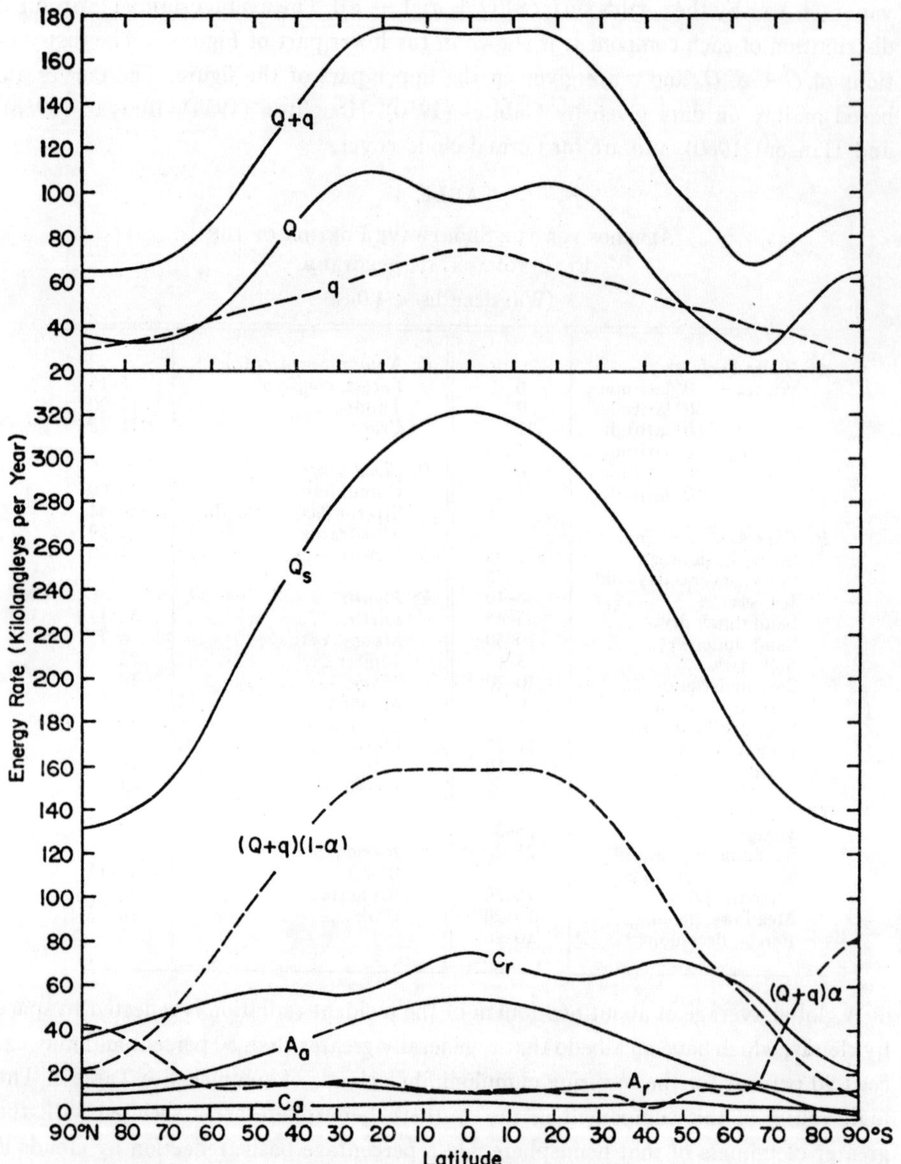

FIG. 7.—The average annual latitudinal distribution of the disposition of solar radiation in the earth-atmosphere system.

A relatively small portion of the incident solar radiation, averaging about 6 percent of Q_s, is scattered back to space by the constituents of the atmosphere, mainly air molecules, dust, and water vapor. A somewhat greater amount, exceeding 20 percent of Q_s, is scattered downward and reaches the surface as diffuse shortwave sky radiation. As indicated in Figure 6, this scattering is selective in that the shorter (blue) wavelengths are scattered more readily than the longer (red) wavelengths, at least when the scattering particles are of the sizes normally found in a cloudless atmosphere. If the circumference of the scattering particles is less than about one-tenth of the wavelength of the incident radiation, the scattering coefficient is inversely proportional to the fourth power of the wavelength. This is frequently referred to as Rayleigh scattering and is the primary cause for the blue sky. For larger size particles with circumferences of more than thirty times the wavelength of the incident radiation the scattering is independent of wavelength or "white."

The atmosphere does not absorb solar radiation well. Less than 3 percent of Q_s is absorbed by clouds. Individual cloud systems certainly do not absorb more than 20 percent of the incident radiation, the average value probably lying closer to 10 percent (Fritz, 1951).

The clear atmosphere absorbs better than do clouds, especially in the tropics where water vapor is abundant. But even here absorption equals only 17 percent of the radiation at the top of the atmosphere. The primary absorption at the shorter wavelengths, that is, below 0.3μ, is due to ozone and molecular oxygen. Essentially none of the radiation in this region reaches the ground (Fig. 6). At wavelengths longer than 0.7μ the solar beam is strongly depleted by water vapor and carbon dioxide absorption.

So far 69 percent of the solar radiation incident on the top of the atmosphere has been accounted for: 30 percent (78 kly year^{-1}) is reflected and scattered to space by clouds (24 percent) and atmospheric constituents (6 percent); 17 percent (45 kly year^{-1}) is absorbed by clouds (3 percent) and atmospheric constituents (14 percent); and 22 percent (58 kly year^{-1}) reaches the surface as diffuse sky radiation. The remainder (31 percent) must reach the surface as direct-beam solar radiation. Thus, on the average, slightly more than half of the solar radiation intercepted by the earth eventually reaches the surface.

Because of its drier atmosphere, the northern hemisphere receives about 4.5 percent more solar radiation than the southern hemisphere at the surface. The secondary maximum at the South Pole, shown in the top part of Figure 7, is due to the highly transparent Antarctic atmosphere. On a percentage basis, solar radiation reaching the ground is greatest at the South Pole and in the subtropics of the northern hemisphere; it is least in the higher middle latitudes of the southern hemisphere, where an extensive cloud cover greatly reduces the amount of direct-beam solar radiation at the surface.

Figures 8, 9, and 10 show the longitudinal and seasonal variation of the total solar radiation reaching the ground $(Q + q)$. The annual distribution over the globe according to Budyko (1963*a*) is shown in Figure 8. Maxima in excess of 200 kly year^{-1} are found over the world's major deserts, the Sahara and Libyan deserts of North Africa, the Arabian Desert of Saudi Arabia, the Thar Desert of Pakistan, and the Mohave and Sonoran deserts of North America. In these regions as much as 80 percent of the solar radiation incident on the top of the atmosphere during the year reaches the ground. Values of less than 100 kly year^{-1} occur poleward of 40° over the oceans, poleward of 50° over the continents, and in the tropical jungle lying between the mouth of the Niger River and the equator on the west coast of Africa.

The average solar radiation reaching the ground in the United States in January and July is shown in Figures 9 and 10, respectively. In January there is an irregular but gradual decrease with latitude from a high of 350 ly day^{-1} in the Sonoran Desert and in southern Florida to a low of less than 100 ly day^{-1} in the Pacific Northwest and Canada. The pattern is much more irregular in July. Values range from a low of less than 500 ly day^{-1} in the Pacific Northwest, the Great Lakes area, and most of New England to a high of 750 ly day^{-1} in the Mojave Desert of California.

Inyokern, California, at an elevation of 744 m in the arid rain shadow on the east side of the southern Sierra Nevada Mountains receives more solar radiation during the year than any other point in the United States for which data are available. The day-by-day variation of $Q + q$ at this station during 1962 is shown in Figure 11. Also shown is the solar radiation received on a horizontal surface at the top of the atmosphere at the latitude of Inyokern (35°39′N) and the envelope of the Inyokern data which represents an estimate of the radiation reaching the ground with perfectly clear skies. During the period from 1951 through 1963 an average of approximately 82 percent of the solar energy incident on the top of the atmosphere at Inyokern reached the surface. This should be compared with the latitudinal average of 57 percent.

All solar radiation incident on the earth's surface is not absorbed there. A certain amount is reflected, depending primarily on the color and composition of the surface. The reflected radiation is lost to space.

Typical albedos for various surfaces are given in Table 4. These albedos are average values for the entire shortwave portion of the electromagnetic spectrum, that is, for wavelengths of less than 4μ. They have been obtained from data given by Budyko (1956), List (1958), Houghton (1954), Geiger (1961), and Kuiper and Middlehurst (1961). The shortwave albedo is highest over fresh snow and above

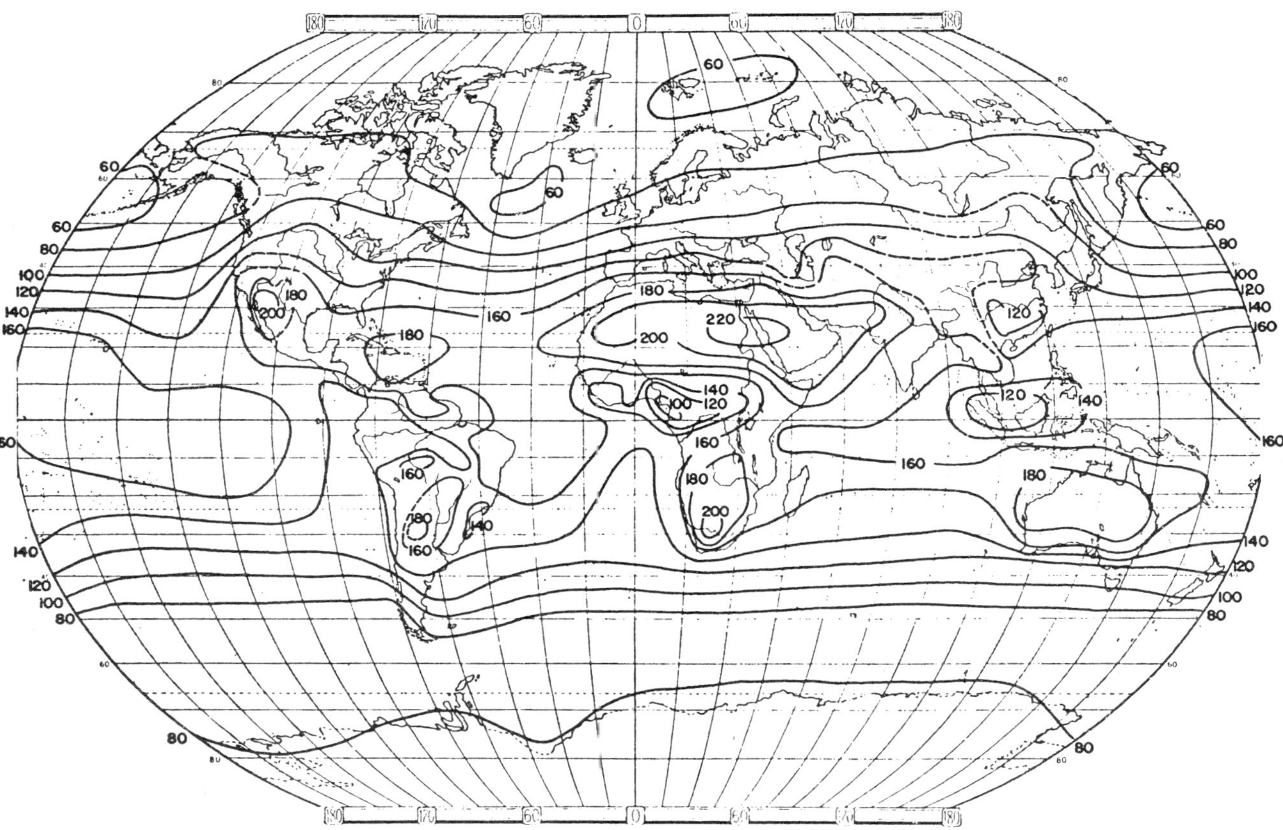

FIG. 8.—The average annual solar radiation on a horizontal surface at the ground. The units are kilolangleys per year. After Budyko (1963*a*). Base map used with the permission of Denoyer-Geppert Co., Chicago.

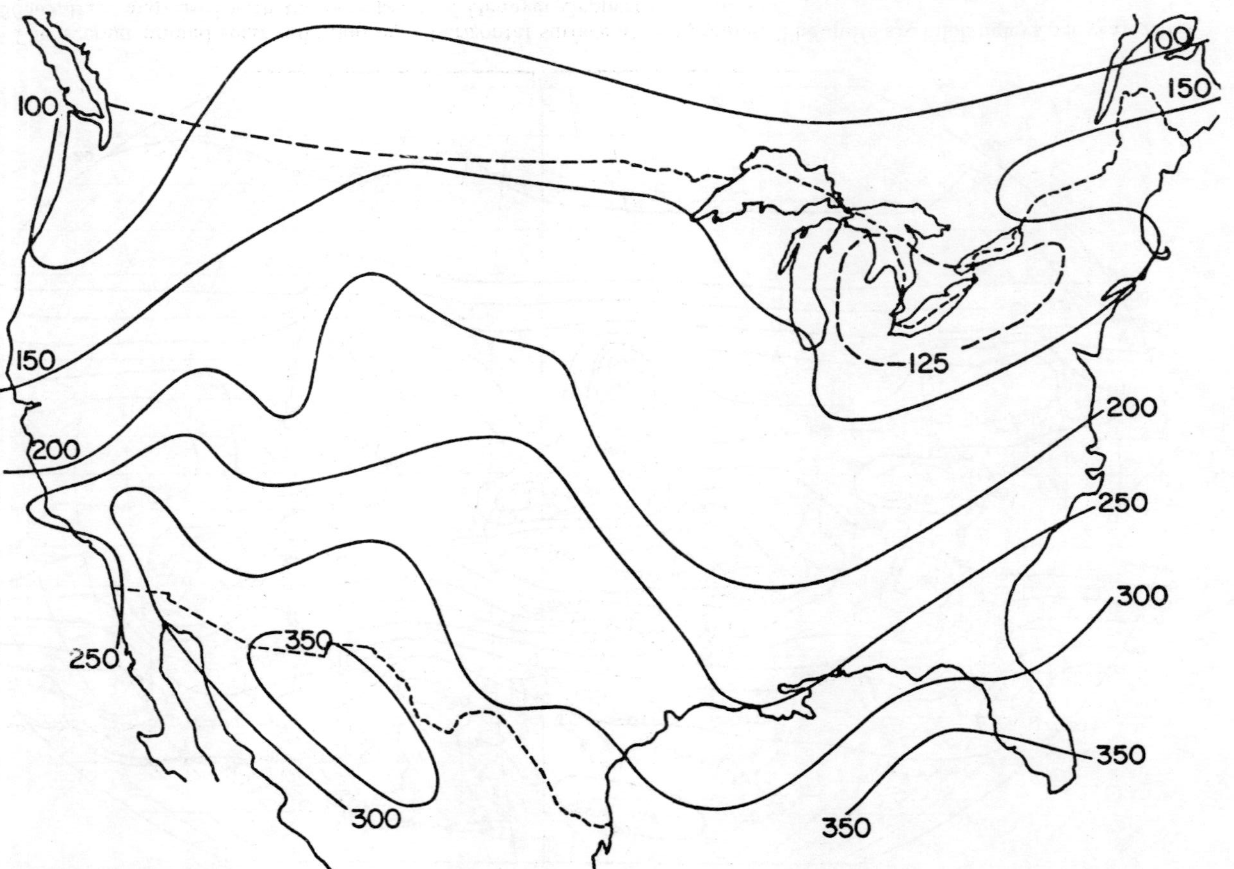

FIG. 9.—The average solar radiation on a horizontal surface at the ground in January in the United States. The units are langleys per day.

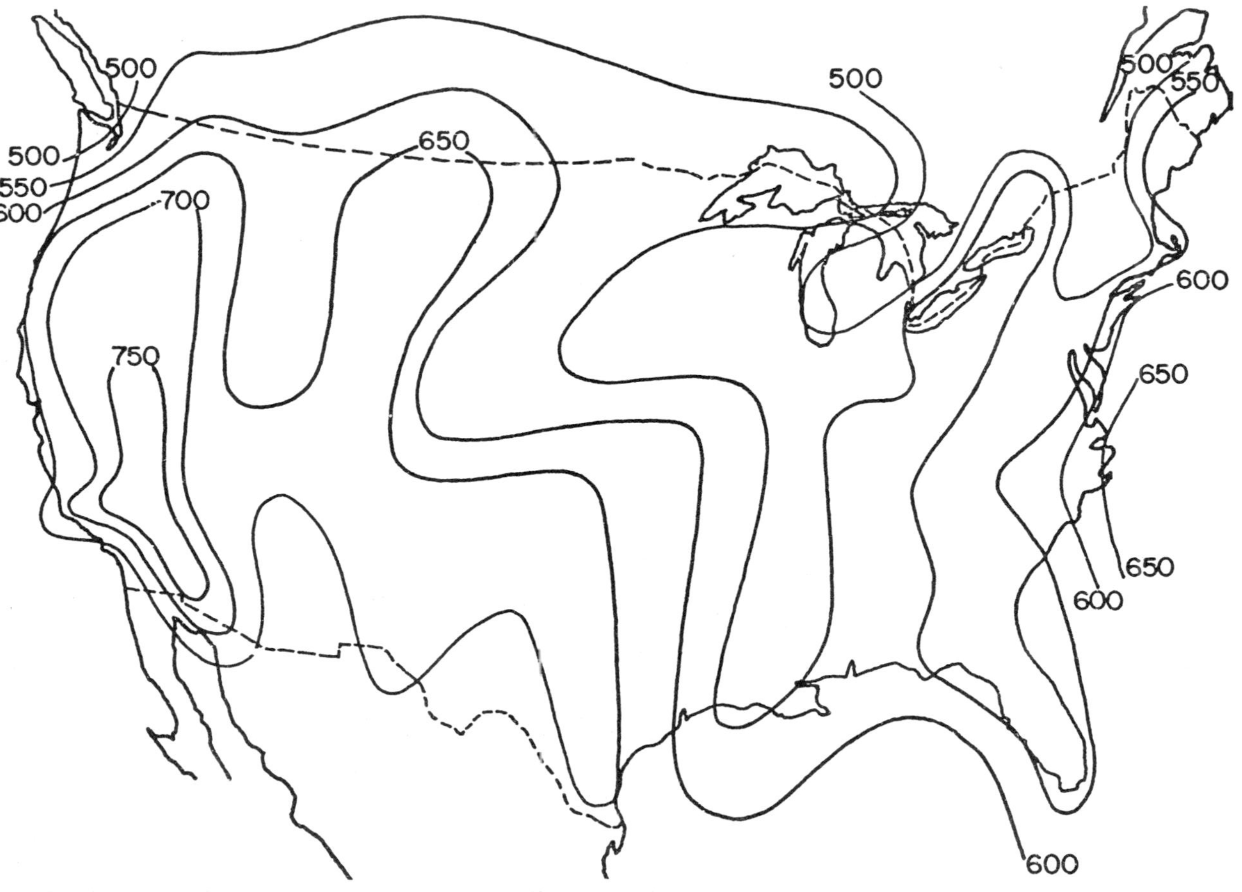

FIG. 10.—The average solar radiation on a horizontal surface at the ground in July in the United States. The units are langleys per day.

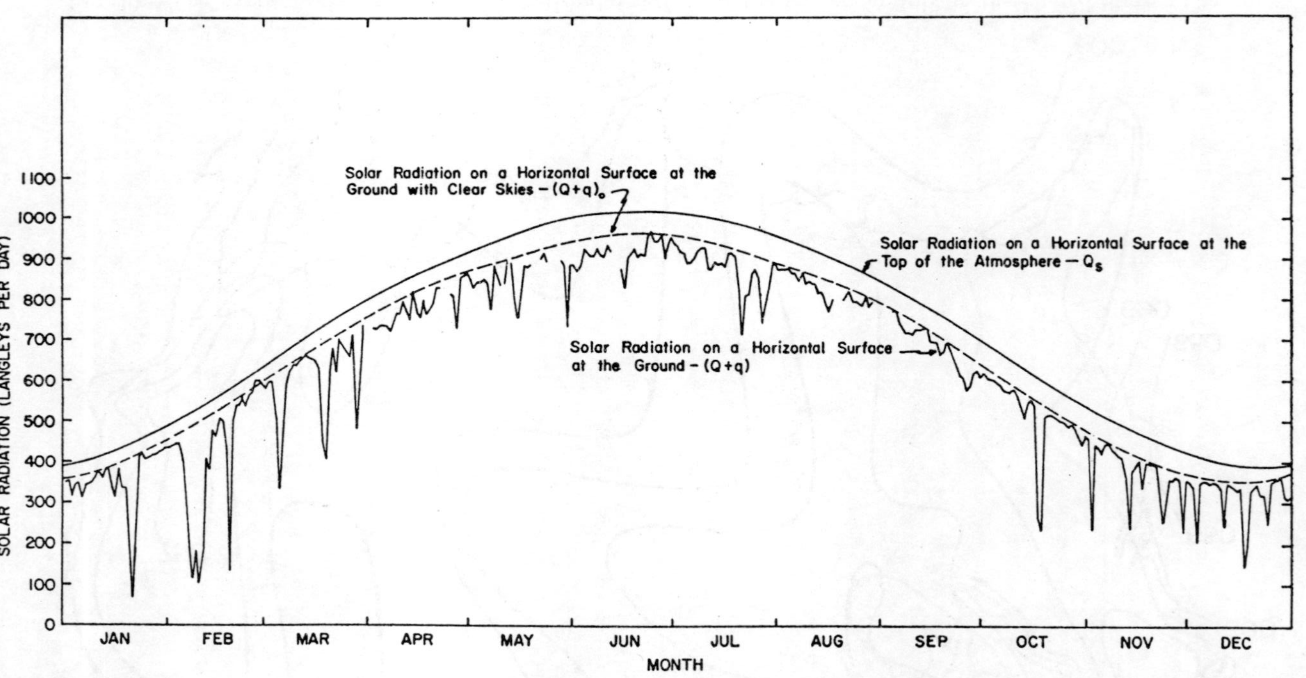

FIG. 11.—Solar radiation on a horizontal surface at the top of the atmosphere, Q_s, at the ground, $Q + q$, and at the ground with clear skies, $(Q + q)_0$, at Inyokern, California (35°39′N, 117°49′W), during 1962.

cumuliform and stratus cloud decks; it is lowest over water and forests. In general, a wet or dark surface has a lower albedo than a dry or light surface. Kung, Bryson, and Lenschow (1964) have recently published maps showing the seasonal distribution of the surface albedo over North America. Their average latitudinal values are reproduced in Table 5. In low latitudes, south of 35°N, the surface albedo varies only slightly from summer to winter, averaging 16 to 19 percent (22 to 26 percent

TABLE 5

LATITUDINAL AND CONTINENTAL MEANS OF SURFACE ALBEDO OVER NORTH AMERICA

[After Kung, Bryson, and Lenschow (1964)]

Latitude Zone (°N)	Prevailing Surface Type	Continental Surface Albedo (percent)				
		January 31			March 31	Summer
		Average Snow Cover	Heavy Snow Cover	Light Snow Cover	Average Snow Cover	
65–70...	Tundra	82.8	82.8	82.7	82.8	16.1
60–65	Tundra and conifer forest	67.3	67.3	58.2	67.3	15.6
55–60		59.1	59.1	54.8	57.7	16.5
50–55	Forest and grassland	50.3	50.3	45.8	48.0	14.6
45–50		46.4	48.9	28.4	37.6	14.8
40–45	Shrubland (Great Basin), cropland (Midwest), and woodland (East)	37.9	50.4	19.0	30.5	15.8
35–40		28.5	40.8	16.0	21.1	16.5
30–35	Desert and shrubland (West); cropland and woodland (East)	19.1	26.2	16.9	17.4	17.2
25–30		17.8	17.8	17.8	17.9	17.9
20–25...	Highlands, desert, rain forest	15.8	15.8	15.8	15.8	15.8
....	Continental mean	43.0	47.4	34.7	39.4	16.0

over the deserts of the Southwest). In higher latitudes the range is much greater, the prevailing snow cover of winter reflecting up to 83 percent of the incident radiation. Especially notable and of great climatic significance is the strong dependence of the surface albedo of middle latitudes (30 to 50°N) on the extent of the snow cover. Even with no change in the incident solar radiation, the energy absorbed by the land surface between 40 and 45°N in a mild winter may exceed that absorbed in a cold, wet winter by more than 60 percent.

The albedo of most surfaces varies both with the wavelength and angle of incidence of the light rays. The dependence of the albedo of wet and dry sand and of pure and polluted ice on the wavelength of the incident radiation is given below. Also shown is the reflectivity of pure dry air. These data are taken from Sauberer (1951) and List (1958). Most types of soil and vegetation have very low albedos in the ultraviolet, increasing in the visible and infrared. Ice, however, appears to have its highest albedo near 0.55μ, with lower values at both shorter and longer wavelengths.

WAVELENGTH (μ)

Surface	Violet (0.4)	Green (0.5)	Orange (0.6)	Red (0.7)	Infrared (0.8)
Dry sand	20	23	29	30	30
Wet sand	10	12	15	16	19
Pure ice	44	54	56	48	32
Polluted ice	24	33	36	31	19
Pure dry air	29	13	6	3	2

According to Budyko (1956) and Geiger (1961), the shortwave albedo over snow-free surfaces is a function of the angle of incidence of the solar radiation, with the highest values occurring near sunrise and sunset. This dependence is most pronounced for plain surfaces, as shown below for sand and moving water (Geiger, 1961).

SOLAR ZENITH ANGLE (DEG)

Surface	40	50	60	70	80	90
Dry sand	35	41	51	63	81	100
Wet sand	26	28	33	43	60	100
Moving water	7	10	16	26	47	100

A similar variation is indicated by some authors for vegetation. Careful measurements by Kuhn and Suomi (1958), however, suggest otherwise. They found that, over prairie grass in Nebraska in summer, the albedo is almost constant from sunrise to sunset, ranging only from 14 to 17 percent.

In any event, cloud cover reduces the diurnal variation by proportionately increasing the diffuse radiation. The latter is almost independent of the solar elevation. The ultraviolet albedo shows very little dependence on the solar altitude, because most of this radiation comes from the sky (as diffuse solar radiation) rather than directly from the sun (see Fig. 6).

The albedo of a plane water surface for direct beam solar radiation is a function of the zenith angle Z and the index of refraction of the water i. It may be computed from Fresnel's formula,

$$\text{Albedo for direct-beam solar radiation} = 50\left[\frac{\sin^2(Z-r)}{\sin^2(Z+r)} + \frac{\tan^2(Z-r)}{\tan^2(Z+r)}\right]$$

where r is the angle of refraction, related to i and Z by Snell's law,

$$i \sin r = \sin Z\,.$$

At a temperature of 20°C, the index of refraction of water relative to air varies from 1.3333 for pure water to 1.340 for sea water with a salinity of 38 parts per mille (35 parts per mille is standard). For normal incidence radiation, Fresnel's formula reduces to

$$\text{Albedo} = 100\left(\frac{i-1}{i+1}\right)^2,$$

which gives an albedo of about 2.4 percent for a plane water surface. If the water is agitated or in motion, the albedo may be as much as three times larger than that obtained from Fresnel's formula, particularly at small zenith angles.

The albedo of water surfaces for diffuse solar radiation has been theoretically determined as about 17.3 percent (Anderson, 1954). Observations under thick overcasts give somewhat lower values (3 to 10 percent). These measurements, however, obviously include some direct-beam radiation, since complete diffusion of solar radiation is attained in nature only with a low overcast cloud cover and very low sun altitudes.

The latitudinal distribution of the solar energy reflected and absorbed annually at the earth's surface is shown in the lower part of Figure 7. The reflected portion, $(Q+q)\alpha$, exceeds the absorbed portion, $(Q+q)(1-\alpha)$, only over the snow fields of the polar regions. At the South Pole almost 60 percent of the energy incident on the top of the atmosphere is reflected back to space from the snow surface and only 10 percent is absorbed. On the other hand, between 35°N and 30°S at least half of the solar energy incident on the top of the atmosphere is absorbed at the surface and only 4 to 6 percent is reflected.

The global disposition of Q_s is outlined in Table 6. The most important point is that of the 169 kly year^{-1} of solar radiation absorbed by the earth-atmosphere system almost 75 percent of this (124 kly year^{-1}) is absorbed at the surface, which, therefore, acts as the main direct source of energy for most of the physiological and thermal processes taking place in the system.

The planetary albedo can be determined from Table 6. This is the percentage of the incident solar radiation that is scattered and reflected back to space by the earth-atmosphere system. This value, about 36 percent, is in good agreement with recent satellite measurements by Astling and Horn (1964).

The energy absorbed by the atmosphere and the earth's surface is directly related to the radiative heating of the two media. Since absorption or emission rates will often be expressed in terms of heating or cooling rates in the following chapters, the appropriate equation will be derived here.

Consider an arbitrary depth Δz of any substance (air, water, or soil) with a density ρ and a horizontal cross section ΔA. Its mass is $\rho\Delta z\Delta A$ and its internal energy is $(\rho\Delta z\Delta A)c\bar{T}_1$, where c is the specific heat of the substance (the amount of heat required to raise the temperature of 1 g 1°C) and $\bar{T}_1$ is its mean temperature. After

TABLE 6

GLOBAL DISPOSITION OF SOLAR RADIATION INCIDENT ON THE TOP OF THE ATMOSPHERE DURING AN AVERAGE YEAR

(kly per year)

Solar energy incident on top of atmosphere (Q_s)	263
Reflected by clouds (C_r)	63
Reflected by molecules, dust, water vapor (A_r)	15
Total reflected by the atmosphere	78
Reflected from the earth's surface [$(Q+q)\alpha$]	16
Total reflected by earth-atmosphere system	94
Absorbed by clouds (C_a)	7
Absorbed by molecules, dust, water vapor (A_a)	38
Total absorbed by the atmosphere	45
Absorbed at the earth's surface [$(Q+q)(1-\alpha)$]	124
Total absorbed by earth-atmosphere system	169

some time Δt, during which heat is added to or taken from the substance, the internal energy will be $(\rho\Delta z\Delta A)c\bar{T}_2$. The change in internal energy during the time interval Δt is then $(\rho\Delta z\Delta A)c(\bar{T}_2 - \bar{T}_1) = (\rho\Delta z\Delta A)c\Delta\bar{T}$, which must equal the net heat gained or lost by the substance. Thus,

$$\text{Net heat gained or lost} = \rho\Delta z\Delta A c\Delta\bar{T} ,$$

or per unit time and cross-sectional area

$$\text{Net heat flux} = \rho c \frac{\Delta\bar{T}}{\Delta t}\Delta z , \qquad (3.11)$$

where $\rho c = C$ is the heat capacity of the substance. If the 124 kly year^{-1} of solar energy absorbed at the earth's surface is completely absorbed in the upper centimeter, then the heating rate in the absence of other compensating processes would be 425°C day^{-1}, if a heat capacity of 0.8 cal cm^{-3} deg^{-1} is assumed. This, of course, does not occur. Almost as fast as the energy is absorbed it is either transferred vertically by longwave radiation and conduction or used to evaporate water.

In applying equation (3.11) to the atmosphere, it is convenient to express the depth of the column in units of pressure. To do this, the hydrostatic equation

$$\Delta p = 10^{-3} \rho g \Delta z$$

is introduced into equation (3.11) giving

$$\text{Net heat flux} = \frac{10^3\ c_p}{g} \frac{\Delta \bar{T}}{\Delta t} \Delta p\,. \tag{3.12}$$

Here, g is the acceleration of gravity, c_p is the specific heat of air at constant pressure (0.24 cal g^{-1} deg^{-1}), and Δp is the pressure thickness of the layer in millibars. With Δp equal to 10^3 mb, the approximate pressure depth of the atmosphere, and g equal to 10^3 cm sec^{-2}, the average heating rate of the atmosphere due to the absorption of 45 kly year^{-1} of solar radiation is about 0.5°C day^{-1}. The latitudinal distribution of this radiative heating is given in Figure 35.

DAYLIGHT ILLUMINATION ON SLOPING SURFACES

The intensity of solar radiation on surfaces sloping in different directions is important in many agricultural and engineering studies. Considering first a vertical wall and representing the intensity of the direct-beam solar radiation on a surface normal to the sun's rays by Q'_n, on a horizontal surface by Q', and on a vertical surface by Q'_v, we can easily see from Figure 12*a* that

$$Q' = Q'_n \cos Z \tag{3.13}$$

and

$$Q'_v = Q'_n \sin Z \cos (a - a') \tag{3.14}$$

where Z is the solar zenith angle, a is the azimuth angle of the sun from south, and a' is the azimuth angle of the normal to the vertical surface from south. Both a and a' are considered negative when directed east of south and positive when directed west of south.

For sloping surfaces other than vertical walls the derivation is more difficult but still straightforward. Let i be the angle between the sloping surface and a horizontal surface and Z' be the angle between the incident solar rays and the perpendicular to

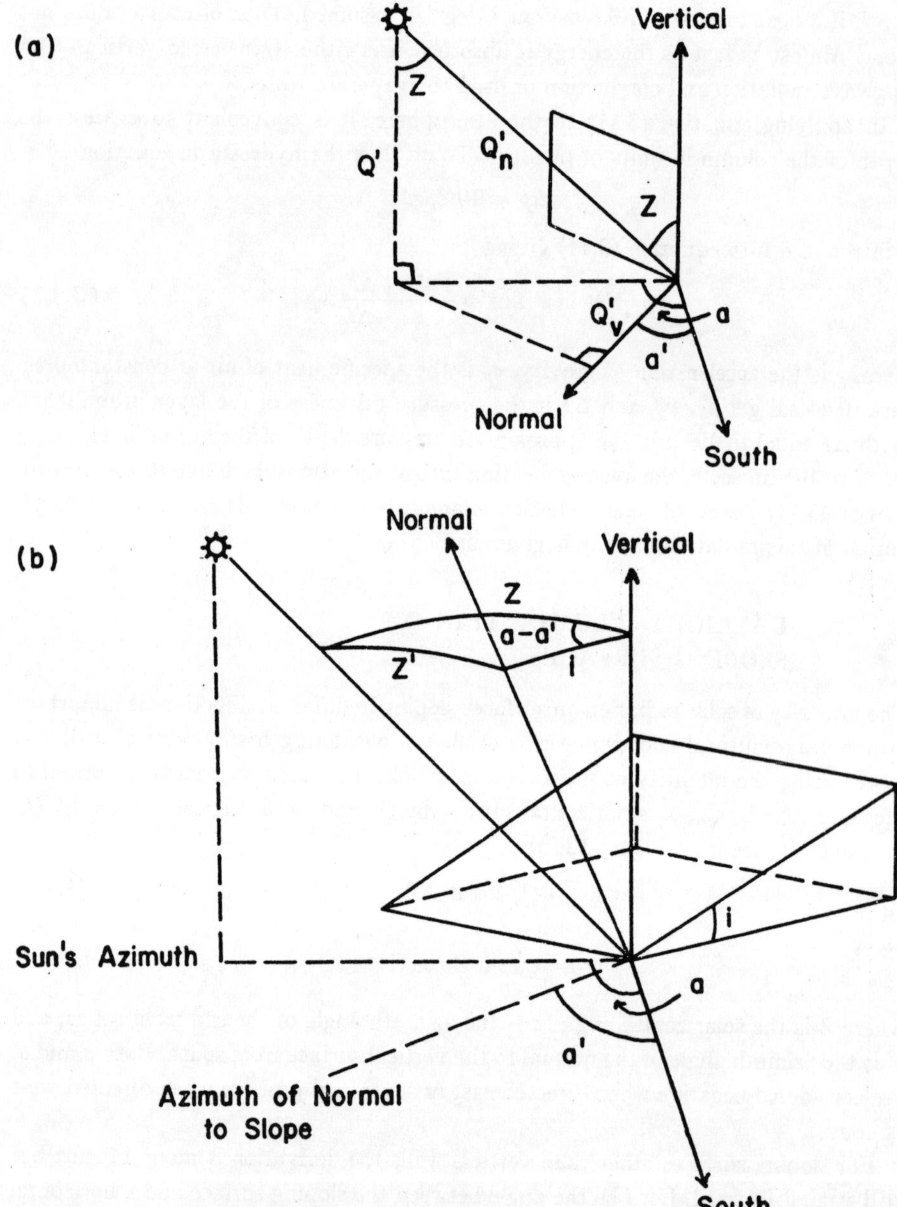

FIG. 12.—Relation of the solar zenith angle Z to the energy incident on (a) a vertical wall and (b) a sloping surface.

the sloping surface. From the geometry of Figure 12*b*, it then follows from the cosine law of spherical trigonometry that

$$Q_i' = Q_n' \cos Z' = Q_n'[\cos Z \cos i + \sin Z \sin i \cos (a - a')] \qquad (3.15)$$

where Q_i' is the instantaneous flux of direct-beam solar radiation on the sloping surface. For a horizontal surface, equation (3.15) reduces to equation (3.13) and, for a vertical wall, to equation (3.14).

The variation of solar intensity with slope, azimuth, and time of year is shown in Figure 13 for 32°N, the latitude of Tucson, Arizona. A similar figure is presented by Geiger (1961) for Trier, Germany (50°N). Data are given for June 21 (top row), March 20 and September 23 (middle row), and December 22 (bottom row) and for north-, east-, south-, and west-facing slopes (the first, second, third, and fourth columns, respectively).

The values shown for a 0° slope are for a horizontal surface and are the same for all slope directions at a given time of year. Symmetry to the noon line exists for the north and south slopes (except for changes in Q_n' due to changes in atmospheric turbidity) but not for east and west slopes.

North slopes.—From March 20 to September 23 the time of sunrise or sunset is the same for all north slopes, because the sun rises in the northeast quadrant and sets in the northwest quadrant. During the winter half-year the sun rises later and sets earlier with increasing slope angle, not shining at all on slopes with an inclination angle greater than the value of $90° - Z$ at noon (58° in March and September and 35° in December at 32°N).

In June there is no direct sunlight at solar noon on north slopes of more than $90° - Z$ (81° at 32°N). The maximum illumination on vertical walls occurs near 0630 and 1730 solar time. For slopes up to about 50° the maximum illumination occurs at noon, with the highest values on a horizontal plane.

East slopes.—The time of peak intensity of the incident radiation on east slopes varies with the slope inclination and the season, occurring earliest on the steeper slopes. The sun's rays are most nearly normal to these slopes in the early morning, but their intensity is greatest at noon. The time of maximum illumination is the resultant of these two factors. At any given time illumination will be greatest on that east slope for which $\tan i = -\tan Z \sin a$. Sunrise occurs at the same time on all east slopes; however, the sun sets earliest on the steeper slopes (at noon on vertical walls).

These conclusions are directly applicable also to west slopes.

South slopes.—From September 21 to March 21, during the winter half-year, the time of sunset or sunrise is the same for all south slopes, because the sun rises in the southeast quadrant and sets in the southwest quadrant. During the summer half-

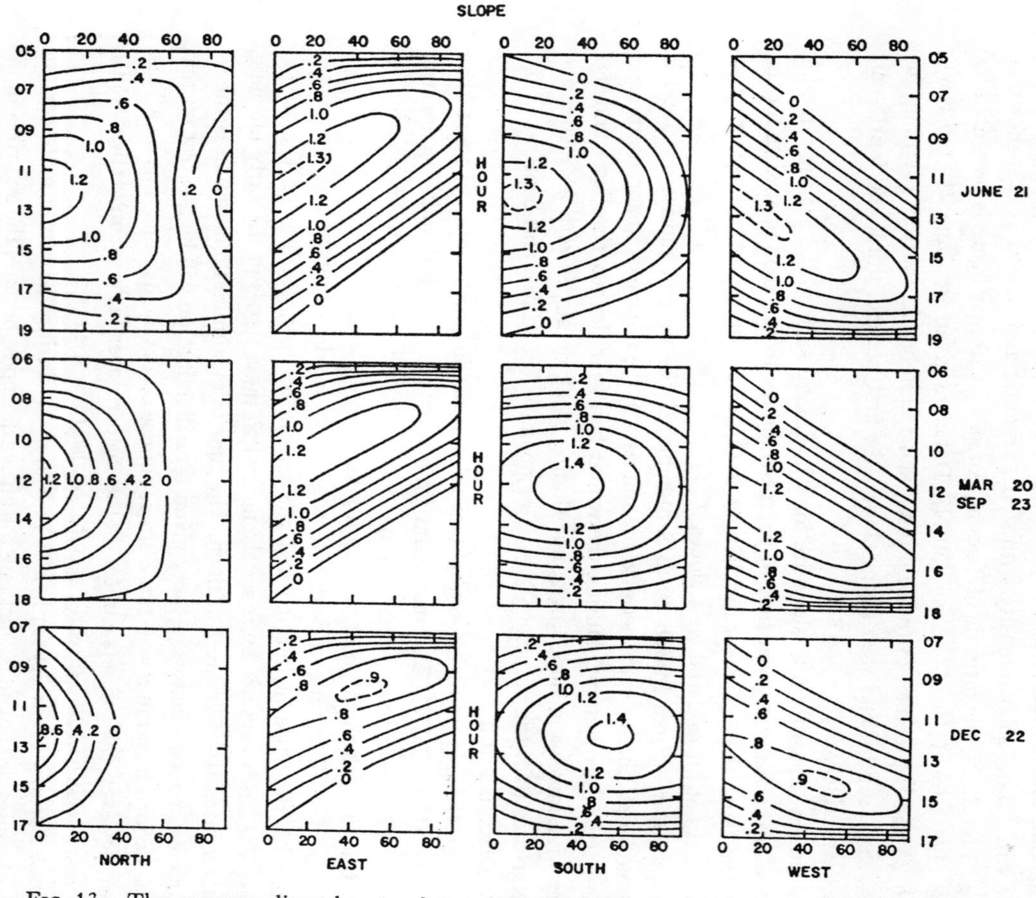

FIG. 13.—The average direct-beam solar radiation incident at the surface at Tucson, Arizona (32°N), as a function of the surface slope and direction, the time of day, and the time of year. The units are langleys per minute.

year, south slopes will be shaded for part of the day. Vertical walls at 32°N will receive sunlight only from 0900 to 1500 on June 21.

On June 21, a south wall in the northern hemisphere intercepts less direct solar radiation than a north wall at all latitudes. The ratio increases from 0 south of 23° to about 0.5 at 30°N and 1.0 at the North Pole.

The solar intensity on any south slope is always a maximum at solar noon. The highest values occur on that slope whose inclination is equal to the zenith angle of the sun, that is, on that slope whose surface is normal to the sun's rays. The intensity of this radiation Q_n' is often greater in winter than in summer in contrast with the annual variation of $Q + q$, the total flux of solar radiation on a horizontal surface, which has a pronounced peak in mid-summer (Fig. 11). At Tucson the noon values of Q_n' range from about 1.45 ly min^{-1} in late winter to 1.30 ly min^{-1} in July and August, when the water vapor content of the atmosphere is very high and depletion of the solar beam by scattering is a maximum.

Equation (3.15) cannot be integrated to give the total direct-beam solar radiation intercepted by a sloping surface during the day, Q_i, because of the unknown dependence of Q_n' on the solar zenith angle. Even if the atmosphere is assumed to be transparent to solar radiation, so Q_n' equals the solar constant, the resulting integrated equation is very complicated and difficult to use. Perhaps the best approach is to perform a graphical integration using equation (3.15) and figures similar to Figure 13. Lee (1963) considers this topic in more detail.

PENETRATION OF SOLAR RADIATION INTO SOIL, WATER, AND ICE

The solar radiation incident on the earth's surface may not be immediately absorbed or reflected but may penetrate to considerable depths depending on the nature of the surface. For most soils the depth of penetration is very small but nevertheless important for processes of soil formation and plant growth. It depends on two factors, the wavelength of the incident light and the size of the soil particles.

Most soils and vegetation readily absorb energy in the ultraviolet portion of the electromagnetic spectrum and scatter and reflect it at longer wavelengths. For this reason, with increasing depth in the soil and as far as sunlight penetrates, proportionately less of the radiation is in the short wavelengths (blue) and more in the longer wavelengths (red). Sauberer (1951), for example, found that blue light (0.47μ) is reduced to 0.01 of its surface intensity at a depth of 1.8 mm in wet sand. At the same depth, orange light is reduced to 0.1 of its surface intensity.

The effect of particle size on the depth of penetration of light in soil has been studied by Baumgartner (1953) with selenium photocells and an artificial light source whose spectrum was not quite the same as that of solar radiation. His results for various grades of dry quartz sand are summarized below. The light intensity is arbitrarily

Depth (mm)	Grain Size (mm)			
	0.2–0.5	0.5–1.0	1.0–2.0	4.0–6.0
0.0	100.0	100.0	100.0	100.0
0.5	35.2	71.8	76.0	95.0
1.0	10.6	54.0	50.5	89.3
1.5	2.5	32.2	40.2	82.3
2.0	0.4	5.7	30.0	74.7
3.0	0.0	0.6	18.0	59.5
5.0	......	0.0	7.3	33.5
10.0	......	......	0.3	5.5
15.0	......	......	0.0	0.6
20.0	......	......	......	0.2

taken as 100 at the surface. In general, the larger the grain size the deeper the radiation will penetrate into the soil. Only for the coarsest soils, however, will significant amounts of radiation reach depths of more than a few millimeters.

For pure water 55 percent of the incident solar radiation penetrates to a depth of 10 cm and 18 percent to a depth of 10 m. The penetration in pure water is very strongly wavelength dependent, as shown in the table below, derived from theoretical calculations by Schmidt (1908) summarized by Dorsey (1940). In contrast to soils, the penetration is greatest at the shortest wavelengths: 72.6 percent of the radiation between 0.2 and 0.6μ reaches 10 m. The penetration is poorest at the longer wavelengths with practically all radiation at wavelengths greater than 1.2μ being absorbed in the upper few centimeters of water.

Depth	Wavelength (microns)			
	0.2–0.6	0.6–0.9	0.9–1.2	1.2–3.0
0.00	100.0	100.0	100.0	100.0
0.01 mm	100.0	100.0	100.0	97.2
0.1 mm	100.0	100.0	99.6	79.0
1.0 mm	100.0	99.8	96.2	40.7
1.0 cm	100.0	98.2	68.7	7.6
10.0 cm	99.7	84.8	4.6	0.0
1.0 m	96.8	36.0	0.0	......
10.0 m	72.6	2.6	......	......
100.0 m	5.9	0.0	......	......

The transmission of sunlight by natural coastal and inland waters varies greatly, depending mainly on the turbidity (detritus, sand, and soil content) and the plankton growth. In general, the more turbid the water the smaller the amount of radiation that can penetrate to a given depth and the higher the albedo. Simultaneously, the wavelength of maximum penetration shifts from the blue for pure water to the red for very muddy water. Geiger (1961) quotes a maximum transmittance per meter of 48 percent at 0.45μ for water in a swamp pond.

Snow and ice are intermediate between soil and water in their ability to transmit solar radiation. Some representative data, taken from Dorsey (1940) and Geiger (1961), are given in the following table. The transmissivity of snow decreases with

Depth (cm)	Wet Snow	Dry Snow	Glacier Ice
2.5	...	...	86.7
5	8.0	...	75.2
10	2.4	18.5	56.6
15	1.1	5.5	42.5
20	...	3.2	32.0
25	...	2.2	24.1
40	...	1.2	10.2
60	...	0.6	3.0

increasing water content. As little as 2 percent or as much as 20 percent of the incident radiation can penetrate to a depth of 10 cm. Thin layers of glacier ice, of the order of 10 to 15 cm thick, transmit as much radiation as pure water. Ten percent of the incident light reaches 40 cm. The dependence of the penetration depth on the wavelength of the incident light is about the same for snow and ice and very similar to that for pure water. Penetration is greatest in the visible between 0.4 and 0.5μ and decreases toward shorter and longer wavelengths.

4 / INFRARED RADIATION

The surface of the earth, when heated by the absorption of solar radiation, becomes a source of longwave radiation. Because the average temperature of the earth's surface is about 285°K, most of the radiation is emitted in the infrared spectral range from 4 to 50μ, with a peak near 10μ, as indicated by the Wien displacement law, equation (3.9). (See Fig. 6.) This radiation is called terrestrial radiation, because it is emitted by the earth's surface, or called nocturnal radiation, because it is the major radiative source of energy at night. Both terms, however, are misleading—the first because the atmospheric constituents also radiate energy in the infrared wavelengths and the second because infrared radiation occurs during the day as well as at night.

The earth's surface is commonly assumed to emit and absorb energy as a gray body in the infrared region; that is, as a body for which the Stefan-Boltzmann law, equation (3.8), takes the form

$$\text{Terrestrial flux} = I_{\uparrow} = \epsilon\sigma T^4 \tag{4.1}$$

where the constant of proportionality ϵ is defined as the infrared emissivity or, equivalently, the infrared absorptivity (one minus the infrared albedo). The emissivity of a black body is 1.

Typical infrared emissivities, expressed in percent, for various surfaces are given in Table 7. Most of these data have been taken from Geiger (1961) and Brooks (1959). Practically all surfaces, including snow, have emissivities of between 90 and 95 percent, corresponding to an infrared albedo of 5 to 10 percent. Recent satellite measurements, analyzed by Buettner and Kern (1963), suggest that these values may be too high. They obtained an emissivity of between 69 and 91 percent for the spectral range from 8 to 12μ over the Libyan and Egyptian deserts. As Wark, Yamamoto, and Lienesch (1962) point out, however, there are numerous hazards in estimating surface temperatures and, hence, emissivities from satellite measurements alone.

The emissivity of most leaves and plants reaches a secondary minimum of 60 to 80 percent at 0.55μ in the visible, near the peak wavelength of the solar spectrum. It reaches a primary minimum of 10 to 50 percent in the infrared between 0.74 and 1.10μ (Gates *et al.*, 1965). The higher values apply to desert succulents. According to Gates (1963), the relatively low emissivity of plants in that part of the spectrum where solar radiation is a maximum keeps them from getting too warm in the midday sun. Further, high emissivities, exceeding 90 percent, at longer wavelengths where solar radiation is insignificant, provide an efficient cooling mechanism for the leaf surface.

TABLE 7

INFRARED EMISSIVITIES

(Percent)

A. Water and Soil Surfaces		*C. Vegetation*	
Water	92–96	Alfalfa, dark green	95
Snow, fresh fallen	82–99.5	Oak leaves	91–95
Snow, ice granules	89	Leaves and plants	
Ice	96	0.8μ	5–53
Soil, frozen	93–94	1.0μ	5–60
Sand, dry playa	84	2.4μ	70–97
Sand, dry light	89–90	10.0μ	97–98
Sand, wet	95		
Gravel, coarse	91–92	*D. Miscellaneous*	
Limestone, light gray	91–92	Paper, white	89–95
Concrete, dry	71–88	Glass pane	87–94
Ground, moist, bare	95–98	Bricks, red	92
Ground, dry plowed	90	Plaster, white	91
		Wood, planed oak	90
B. Natural Surfaces		Paint, white	91–95
Desert	90–91	Paint, black	88–95
Grass, high dry	90	Paint, aluminum	43–55
Fields and shrubs	90	Aluminum foil	1–5
Oak woodland	90	Iron, galvanized	13–28
Pine forest	90	Silver, highly polished	2
		Skin, human	95

The average annual latitudinal distribution of infrared terrestrial radiation $I_\uparrow$, which closely follows the latitudinal distribution of surface temperature, is shown in the upper part of Figure 14. The global value, 258 kly year^{-1}, is only slightly less than the average shortwave solar radiation incident on the top of the atmosphere, which is 263 kly year^{-1}.

Although the atmosphere is nearly transparent to shortwave radiation, it readily absorbs terrestrial radiation, the principal absorbers being water vapor (5.3 to 7.7μ and beyond 20μ), ozone (9.4 to 9.8μ). carbon dioxide (13.1 to 16.9μ), and clouds (all wavelengths). Only about 9 percent of the terrestrial radiation escapes directly to space, mainly in the "atmospheric window" (8.5 to 11.0μ); the rest is absorbed by the atmosphere (Fig. 6). The atmosphere, in turn, reradiates the absorbed terrestrial

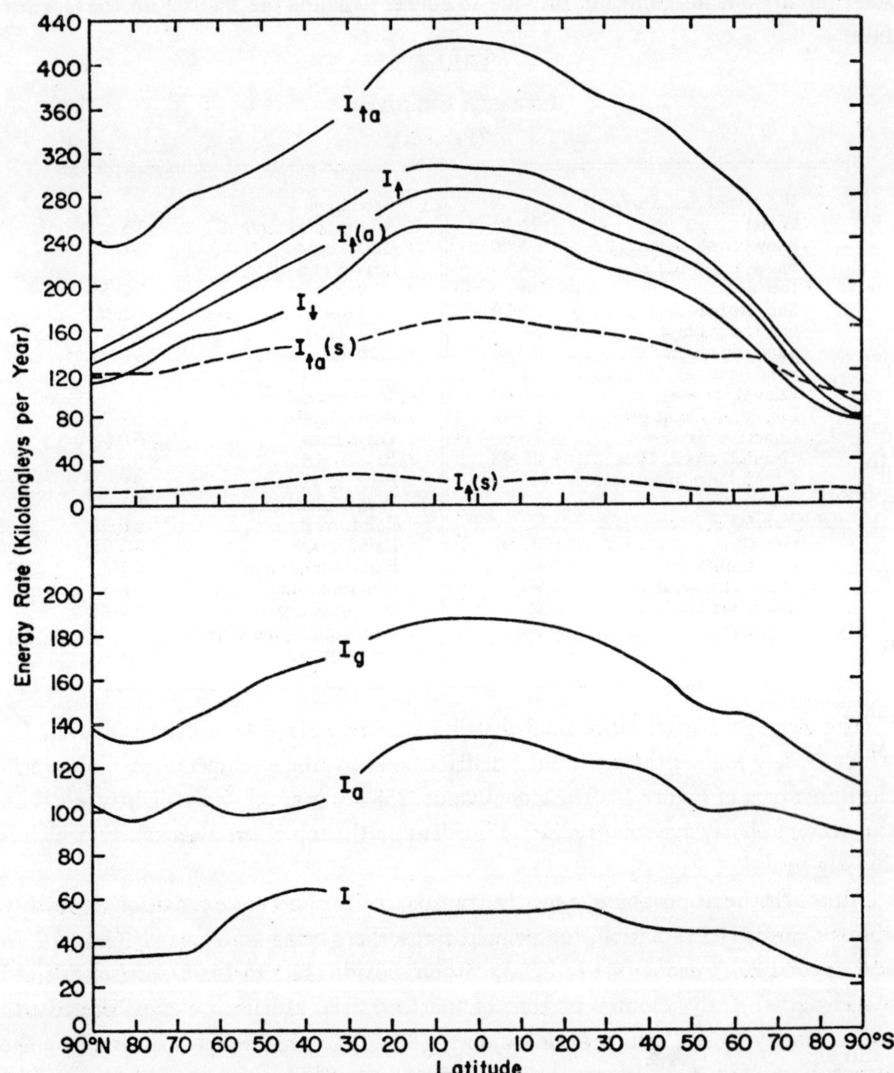

FIG. 14.—The average annual latitudinal distribution of the disposition of infrared radiation in the earth-atmosphere system.

radiation, partly to space and partly back to the surface (counter radiation). Thus the net or effective outgoing radiation from the surface is reduced considerably from what would be observed if the atmosphere were perfectly translucent.

In the absence of atmospheric counter radiation the earth's surface would be 30 to 40°C cooler than it is (Fleagle and Businger, 1963). The same type of surface heating occurs in a greenhouse in which the net outgoing infrared radiation is reduced to almost zero by the opaque glass cover. Hence, the term "greenhouse effect" is often applied to the atmospheric phenomenon.

The principles of the preceding paragraphs may be represented symbolically by the equations

$$I_{\uparrow} = I_{\uparrow}(a) + I_{\uparrow}(s)$$

$$I_{\uparrow a} = I_{\downarrow} + I_{\uparrow a}(s)$$

and

$$I = I_{\uparrow} - I_{\downarrow} \tag{4.2}$$

in which $I_{\uparrow}(a)$ and $I_{\uparrow}(s)$ are, respectively, the infrared radiation from the earth's surface absorbed by the atmosphere, and that lost to space; $I_{\uparrow a}$ is the infrared radiation from the atmosphere; $I_{\downarrow}$ and $I_{\uparrow a}(s)$ are, respectively, the infrared radiation from the atmosphere absorbed at the ground (counter radiation) and that lost to space; and I is the effective outgoing radiation from the earth's surface. It is also possible to define the effective outgoing radiation from the atmosphere I_a and from the earth-atmosphere system I_g as the difference between emitted and absorbed infrared energy, that is,

$$I_a = I_{\uparrow a} - I_{\uparrow}(a)$$

$$I_g = I_{\uparrow}(s) + I_{\uparrow a}(s) = I + I_a\,. \tag{4.3}$$

The mean annual latitudinal distribution of each component of these equations is shown in Figure 14. Most data for this figure were taken from Houghton (1954), Gabites (1950), Budyko (1956), and Hanson (1960).

Fluctuations in the amount of infrared radiation passing unabsorbed to space from the ground, $I_{\uparrow}(s)$, are associated mainly with variations in the surface temperature and in the amount of water vapor and number of clouds in the atmosphere. Cloud droplets, unlike water vapor, absorb and emit energy at all wavelengths, including those of the atmospheric window. Hence, the largest ground losses to space are found in the dry subtropics of the northern hemisphere where cloud cover is at a minimum.

The energy emitted and absorbed by the atmosphere depends partly on the amount of water vapor, carbon dioxide, and ozone present. These gases are selective absorbers and emitters in that they absorb and emit energy almost completely at some wavelengths and practically not at all at others. Their emissivities, therefore, vary with wavelength. It is, nevertheless, still possible to define an over-all emissivity as the percentage of the total black body intensity emitted or absorbed by a layer or column of a gas over the whole spectral range.

The emissivity, thus defined, varies for a given amount of gas and will normally be greater the lower the gas is found in the atmosphere. The absorption bands of water vapor, carbon dioxide, and ozone are made up of many individual lines about 0.01μ wide. The width of the lines is directly proportional to the number of impacts experienced by the gas molecules per unit time and, hence, is proportional to the total air pressure p. Infrared radiation from the "wing" regions originating at low levels (high pressure) is able to escape almost unhindered to space, but emission near the centers of the lines is rapidly absorbed by the upper layers.

To take this effect into account, the true depth of a given gas is replaced by its corrected optical depth or thickness. The corrected optical depth is the true depth multiplied by the ratio of the mean pressure $\bar{p}$ of the layer in which the gas is found to the standard sea level pressure p_o (1,013.25 mb). For water vapor, the true depth is usually measured in centimeters of precipitable water vapor w_a; so the corrected optical thickness for water vapor w_a' is given by

$$w_a' = w_a(\bar{p}/p_o) .$$

For carbon dioxide and ozone, the true depth is measured as the length in centimeters of a column of the pure gas at standard temperature (288°K) and pressure. In these units the total atmospheric content of the three gases averages 1 to 4 cm for water vapor, about 250 cm for carbon dioxide, and 1 to 4 mm for ozone.

Typical emissivities for perfectly diffuse radiation, usually called slab emissivities, are given for water vapor, carbon dioxide, and ozone in Table 8*a*. The values are derived from data presented by Elsasser and Culbertson (1960) and are valid for a temperature of 20°C. In many atmospheric radiation problems the effects of carbon dioxide and ozone are neglected, mainly because of their very low emissivities, and only emission and absorption by water vapor are considered. Ozone, however, is important in the radiation balance of the upper atmosphere, between 20 and 30 km, where its concentrations are greatest.

The emissivities of all three gases vary slightly with temperature and are usually highest when a major absorption band for a given gas lies within the wavelength span of maximum black body emission, which is temperature dependent [see eq. (3.9)]. The variation of emissivity with temperature is shown in Table 8*b* for a column of 2 cm of water vapor, a column of 200 cm of carbon dioxide, and a column of 0.2 cm of ozone.

TABLE 8*a*

SLAB EMISSIVITIES (PERCENT) AS A FUNCTION OF PATH LENGTH (CM) FOR WATER VAPOR, CARBON DIOXIDE, AND OZONE[1]

(After Elsasser and Culbertson, 1960)

Path Length[2]	Water Vapor	Carbon Dioxide	Ozone
0.00001	1.89	....	....
0.00002	2.49	....	....
0.00005	3.78	....	....
0.0001	5.11	0.11	0.22
0.0002	6.83	0.15	0.35
0.0005	9.76	0.22	0.61
0.001	12.45	0.31	0.90
0.002	15.51	0.43	1.30
0.005	19.96	0.67	2.02
0.01	23.43	0.94	2.70
0.02	27.02	1.32	3.47
0.05	31.98	2.12	4.56
0.1	35.92	3.01	5.37
0.2	40.18	4.21	6.13
0.5	46.71	6.15	7.06
1.0	52.81	7.70	7.66
2.0	60.26	9.26	8.14
5.0	72.39	11.30	8.60
10.0	81.38	12.80	8.82
20.0	88.04	14.27	....
50.0	91.44	16.19	....
100.0	....	17.62	....
200.0	....	19.01	....
500.0	....	20.83	....
1,000.0	....	22.17	....
2,000.0	....	23.48	....

[1] Values are for 20°C.

[2] Pressure-corrected centimeters of precipitable water for water vapor, pressure-corrected length in centimeters of a column of the pure gas at standard temperature and pressure for carbon dioxide and ozone.

TABLE 8*b*

SLAB EMISSIVITIES (PERCENT) AS A FUNCTION OF TEMPERATURE FOR WATER VAPOR, CARBON DIOXIDE, AND OZONE

(After Elsasser and Culbertson, 1960)

Temperature (° C)	Water Vapor (Path Length =2 cm)	Carbon Dioxide (Path Length =200 cm)	Ozone (Path Length =0.2 cm)
−80	70.00	15.91	2.13
−70	67.10	16.71	2.58
−60	64.85	17.38	3.04
−50	63.14	17.94	3.50
−40	61.87	18.38	3.95
−30	60.97	18.71	4.38
−20	60.38	18.93	4.79
−10	60.05	19.07	5.17
0	59.94	19.12	5.52
10	60.02	19.09	5.84
20	60.26	19.01	6.13
30	60.63	18.87	6.38
40	61.12	18.68	6.61

Except near the poles, the atmosphere emits more energy to the earth's surface than it does to space, primarily because the downward flux usually originates in a lower and warmer layer in the atmosphere than does the upward flux. Most counter radiation comes from the lowest 100 m of the atmosphere. The data below are from a sounding taken at 0635 CST on August 31, 1953, at O'Neill, Nebraska (Lettau and Davidson, 1957). In this case more than 25 percent of the counter radiation originated below 2 m. In general, with clear skies about 90 percent of the counter radiation comes from the lowest 800 to 1,600 m of the atmosphere.

Near the poles the upward flux exceeds the downward flux partly because, in the presence of the strong surface temperature inversion often found in these regions, the downward flux is emitted at lower temperatures than is the upward flux.

TOTAL COUNTER RADIATION

Percent	Originating below
9.3	0.1 m*
15.9	0.4
20.3	0.8
25.8	2.0
35.0	6.0
44.6	20.0
58.9	100.0
74.6	400.0
84.8	1,000.0
98.5	4,000.0

As shown in Figure 14, both the earth's surface and the atmosphere lose more infrared energy than they gain at all latitudes. Hence, both are cooling radiatively in the long wavelengths. Net losses from the surface are greatest in the lower middle latitudes and subtropics of the northern hemisphere. In the atmosphere, losses are greatest in the tropics.

The average annual global disposition of infrared radiation for the earth-atmosphere system is summarized in Table 9. The net infrared loss from the atmosphere, 117 kly year^{-1}, is, from equation (3.12), equivalent to a cooling rate of 1.3°C day^{-1}, which is greater than the corresponding heating rate of 0.5°C day^{-1} owing to the absorption of solar radiation. On the other hand, the net infrared loss from the earth's surface, 52 kly year^{-1}, is considerably less than the 124 kly year^{-1} gained by the absorption of solar radiation. Hence, the atmosphere is cooling and the earth's surface warming radiatively. The whole system is in equilibrium, since just as much energy, 169 kly year^{-1}, is emitted to space in the long wavelengths as is gained from the sun in the short wavelengths. This figure is in excellent agreement with the value of 170 kly year^{-1} estimated by Astling and Horn (1964) from satellite measurements. The significance of these results is discussed in chapter 5.

TABLE 9

GLOBAL DISPOSITION OF INFRARED RADIATION IN THE EARTH-ATMOSPHERE SYSTEM DURING AN AVERAGE YEAR

(kly per year)

Infrared radiation emitted by the earth's surface ($I\uparrow$)	258
Lost to space [$I\uparrow(s)$]	20
Absorbed by the atmosphere [$I\uparrow(a)$]	238
Infrared radiation emitted by the atmosphere ($I\uparrow_a$)	355
Lost to space [$I\uparrow_a(s)$]	149
Absorbed by the earth's surface as counter radiation ($I\downarrow$)	206
Effective outgoing radiation from the earth's surface (I)	52
Effective outgoing radiation from the atmosphere (I_a)	117
Effective outgoing radiation from the earth-atmosphere system (I_g)	169

THE EFFECTIVE OUTGOING RADIATION FROM THE EARTH'S SURFACE

The effective outgoing radiation from the earth's surface consists of two basic components, the total longwave energy sent out from surface $I\uparrow$, which is a function of the surface emissivity ϵ and temperature T_s, and the counter radiation from the atmosphere $I\downarrow$ which is a function primarily of the air temperature T_a, the precipitable water vapor w_a, and the cloud cover n. Thus,

$$I = I\uparrow - I\downarrow = f(T_a, T_s, \epsilon, w_a, n) .$$

The components of the effective outgoing radiation can be determined by direct measurement using a pyrgeometer, from a radiation chart, or by empirical equations derived from data provided by the first two methods. Pyrgeometers, which are far from perfected instruments, will be discussed in chapter 6. This chapter will deal only with radiation charts and empirical equations.

Radiation charts.—There are a large number of radiation charts available for computing infrared fluxes both in the atmosphere and at the earth's surface. In general, these differ only in relatively minor details, involving the assumptions necessary to make the complex theory of radiative transfer tractable. The chart most commonly used in the United States was developed by Elsasser in 1942 and improved by Elsasser and Culbertson in 1960. In the construction of the earlier model of the chart, Elsasser makes the following assumptions:

1. The absorption lines for water vapor are uniformly spaced with the same intensity and half-width.
2. The portion of the spectrum from 13.1 to 16.9μ is black for all atmospheric layers because of carbon dioxide absorption.
3. The transmission of perfectly diffuse radiation through a layer containing w_a centimeters of precipitable water vapor is equal to the transmission of parallel beam radiation through a layer containing 1.66 w_a centimeters of precipitable water vapor.
4. The pressure effect on the width of the absorption lines of water vapor may be adequately taken into account by multiplying the precipitable water vapor by p/p_o, where p_o = 1,000 mb.

A schematic representation of the Elsasser chart is shown in Figure 15. The chart is so constructed that area is proportional to energy flux, with temperature (T in °C) plotted against the corrected optical depth (w_a' in centimeters). The isotherms are vertical lines. The moisture isopleths radiate from the point $T = -273$°C and range in value from $w_a' = 0$ (actually labeled CO_2) to $w_a' = \infty$. An upper wedge-shaped section takes into account emission and absorption by carbon dioxide. A calibration area accompanying the chart permits the direct transformation of areas, usually measured in arbitrary units with a planimeter, into energy fluxes. To simplify the calculations, a special table gives the energy equivalent (in ly 3 hour^{-1}) of areas bounded by certain isotherms and moisture isopleths. The chart is cut off at -80°C, since the right-hand portion is rarely used and the energy totals can be interpolated from the table.

The corrected optical depth at any atmospheric level j a certain distance above or below some reference level $j = 1$, through which the flux is being determined, is defined as the sum of the optical depths of all air layers between the level j and the reference level. That is,

$$w'_{aj} = \sum_{i=1}^{j-1} \Delta w'_{ai}$$

where

$$\Delta w'_{ai} = (\bar{p}_i/p_o)\,\Delta w_{ai} = 10^{-3}\,\bar{q}_i(\bar{p}_i/p_o)\,\Delta p_i\,. \tag{4.4}$$

$\Delta w'_{ai}$ is the corrected optical depth of the ith atmospheric layer in centimeters, Δw_{ai} is the corresponding amount of precipitable water vapor, $\bar{q}_i$ is the mean specific humidity of the ith layer in grams per kilogram, $\bar{p}_i$ is the mean atmospheric pressure of the ith layer in millibars, and Δp_i is the pressure thickness of the ith layer, also in millibars. By definition, the optical depth is always zero at the reference level.

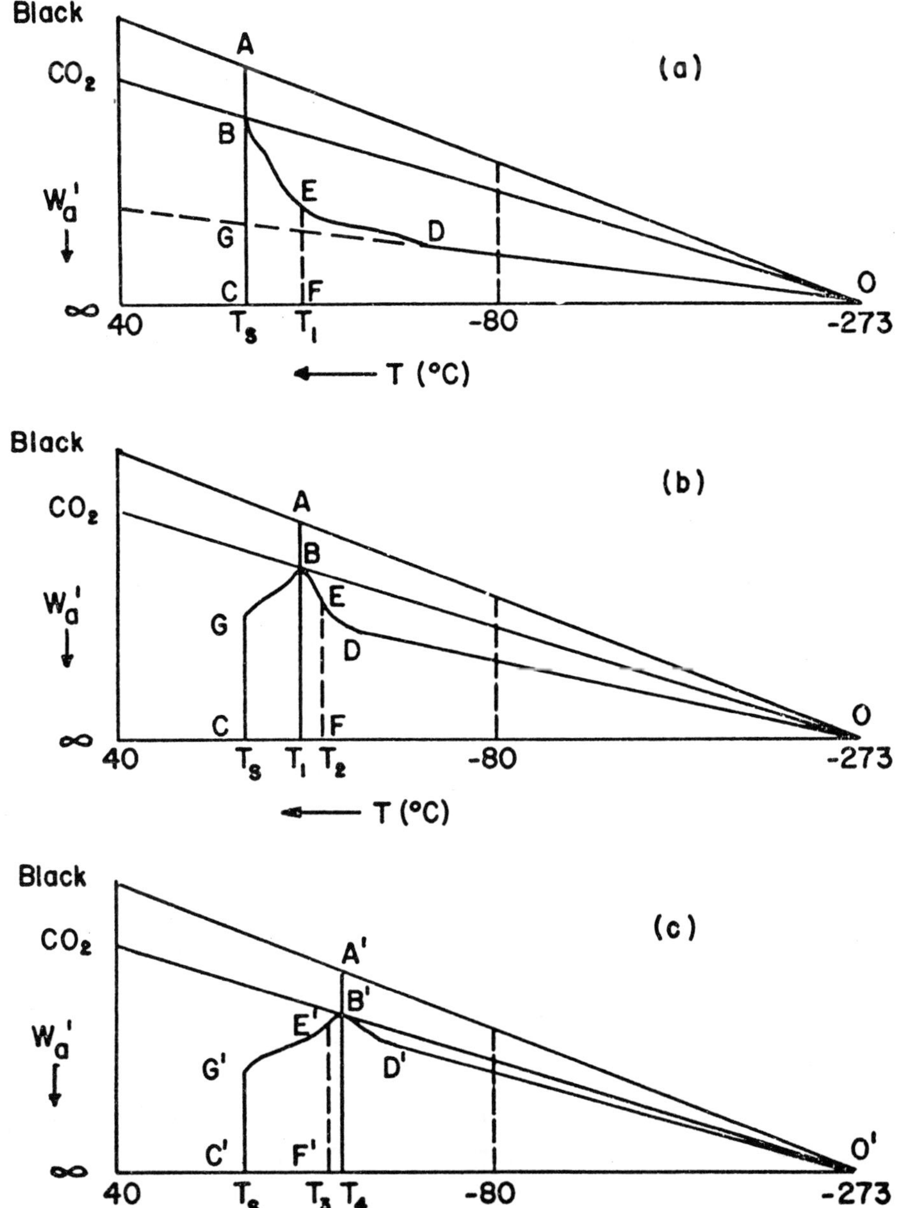

FIG. 15.—The Elsasser radiation chart. Examples are given estimating (*a*) the effective outgoing radiation from the surface, (*b*) the net radiative flux through the 700-mb level, and (*c*) the net radiative flux through the 500-mb level.

The corrected optical depth determined for each level, usually from pressure, temperature, and specific humidity data obtained from radiosonde measurements, is plotted as a function of temperature on the Elsasser chart. A sample of the computations is given in Table 10 for reference levels ($j = 1$) at the surface, 700 mb, and 500 mb. The corresponding plotted data are shown in Figures 15*a*, *b*, and *c*, respectively.

In Figure 15*a* the area to the right of the isotherm T_s is equal to the total black body flux at that temperature. It can represent either the outgoing radiation from the

TABLE 10

ELSASSER CHART COMPUTATIONS

p (mb)	T (° C)	q (g/kg)	q_i (g/kg)	Δp_i (mb)	$\bar{p}_i/p_o$	$\Delta w'_{ai}$ (cm)	w'_{aj} for Reference Level at Surface	700 mb	500 mb
923..........	36	8.4					0	1.33	1.90
			7.5	27	0.91	0.18			
896..........	32	6.6					0.18	1.15	1.72
			7.0	46	.87	.28			
850..........	27	7.4					0.46	0.87	1.44
			7.4	150	.78	.87			
700..........	12	7.3					1.33	0	0.57
			7.0	4	.70	.02			
696..........	11	6.8					1.35	0.02	0.55
			5.4	46	.67	.17			
650..........	9	4.1					1.52	0.19	0.38
			4.6	40	.63	.12			
610..........	5	5.1					1.64	0.31	0.26
			4.2	110	.56	.26			
500..........	− 7	3.3					1.90	0.57	0
			3.0	50	.48	.07			
450..........	−12	2.7					1.97	0.64	0.07
			1.8	20	.44	.02			
430..........	−14	1.0					1.99	0.66	0.09
			0.8	30	0.42	0.01			
400..........	−18	0.7					2.00	0.67	0.10

earth's surface (assuming an emissivity of 1.0), the counter radiation from a cloud layer at temperature T_s, or the counter radiation from an isothermal column of air at temperature T_s containing an infinite amount of water vapor (area *OBCO*) and enough carbon dioxide to provide complete emission between 13.1 and 16.9μ (area *OABO*).

The atmosphere, however, never contains an infinite amount of water vapor and its temperature normally decreases with height. As a result, the counter radiation $I_{\downarrow o}$ absorbed at the ground with clear skies is decreased from $\epsilon\ (OACO)$ to

$$I_{\downarrow o} = \epsilon\ (OABDO)$$

where ϵ is the emissivity of the surface. Part of the decrease, the area *OGCO*, is associated with the presence of only a finite amount of water vapor and the rest, the area *BGDB*, with the decrease of temperature with height.

The earth's surface at a temperature T_s radiates an amount of energy equal to

$$I_{\uparrow} = \epsilon\sigma T_s^4 = \epsilon\,(OACO)\,.$$

The area *ODBCO*, which represents the failure of the atmosphere to radiate as a black body at temperature T_s, must, therefore, be proportional to the effective outgoing radiation with clear skies I_o.

$$I_o = I_{\uparrow} - I_{\downarrow o} = \epsilon\,(ODBCO)\,.$$

Cloud decks are normally treated as black bodies, an assumption criticized by Goody (1964), who points out that the individual particles in a typical water cloud may have an emissivity as low as 0.5 for radiation at 10μ. With this assumption, an overcast whose base is at the temperature level T_1 would increase the counter radiation to

$$I_{\downarrow c} = \epsilon\,(OABEFO)$$

and decrease the effective outgoing radiation to

$$I_c = \epsilon\,(BCFEB)\,.$$

With only a partial overcast at the temperature level T_1, an estimate of the effective outgoing radiation can be obtained from

$$I = (1 - n)\,I_o + nI_c$$

where n is the cloud cover in tenths.

The same concepts apply to estimating the radiative flux through various levels in the atmosphere. For example, in Figure 15*b*, where the reference level is 700 mb, the downward flux is given by the area *OABDO* if the skies are clear and by *OABEFO* if there is an overcast whose base is at the temperature level T_2. The upward flux comes partly from the earth's surface, treated here as a black body, and partly from the water vapor and carbon dioxide below 700 mb. It is given by the area *OABGCO*. This is less than the surface black body radiation, most of which is absorbed below 700 mb and reemitted at a lower temperature than T_s. The net upward flux through the 700-mb level is *ODBGCO* with clear skies and *BGCFEB* with an overcast at the temperature level T_2.

If there is a cloud layer whose top is below the reference level, as, for example, at T_3 in Figure 15*c*, it is almost as though the earth's surface were displaced vertically, the upward flux changing from the area *O'A'B'G'C'O'* to *O'A'B'E'F'O'*. The downward flux is unaffected, being equal to the area *O'A'B'D'O'* in both cases.

The magnitude of infrared warming or cooling in atmospheric layers may be determined from the Elsasser chart. As shown in Fig. 15*b* and *c*, with clear skies, the net upward flux through the 700-mb level is equal to the area *ODBGCO* and through the 500-mb level to the area *O′D′B′G′C′O′*. Planimeter measurements show the latter area to be slightly larger than the former, indicating that more infrared energy is being emitted by the 200-mb layer than is being absorbed. The result should be slight radiative cooling. With a cloud layer between 700 and 500 mb, the net upward flux is equal to the area *BGCFEB* through 700 mb and to *O′D′B′E′F′O′* through 500 mb. Here the latter area is obviously much larger than the former, and cooling rates are correspondingly greater than with no clouds. This pronounced radiative cooling will normally be more than offset by heating due to convection and condensation within the cloud layer.

Other radiation charts which are frequently mentioned in the meteorological literature are those of Robinson (1950, Kew chart), Möller (1951), and Yamamoto (1952).

The Möller chart is very similar to the Elsasser chart, except that total absorption by carbon dioxide is assumed to occur only between 13.5 and 16.5μ, rather than between 13.1 and 16.9μ. As a result, the emissivity of carbon dioxide is reduced from 0.184 to 0.146 at 273°K. Slightly different values of the water vapor absorption line intensities are also used.

The Kew chart is a simplified version of the Elsasser chart in which the main assumption is that the emissivity of perfectly diffuse radiation from a water vapor slab is independent of temperature and a function only of the corrected optical depth. The ordinate is emissivity in percent or the corrected optical depth in centimeters. The abscissa is temperature in °C, the scale being linear in the fourth power of the absolute temperature.

The Yamamoto chart, described in detail by Goody (1964), takes into account the dependence of line intensity on temperature, as well as on pressure. It has essentially the same ordinate and abscissa as the Kew chart. The interaction between water vapor and carbon dioxide is considered with the aid of a series of auxiliary lines. Elsasser and Culbertson (1960) have adopted the same procedure in their recent study.

Empirical equations.—Radiation charts give more information than is usually needed in climatological studies. The computations required are time consuming and laborious, and in many cases the necessary data are not available. Therefore, many simple formulas relating commonly measured meteorological parameters (temperature, vapor pressure, and humidity) to the effective outgoing radiation with clear skies I_o, as measured directly with a pyrgeometer or indirectly with a radiation chart, have been presented. A few of these are discussed below.

1. *The Ångström Equation.* This equation, suggested by Ångström in 1916, is written in numerous different but equivalent forms in the literature; the most common are

$$I_{\downarrow o} = \epsilon\sigma T^4(a_o - b_o 10^{-c_o e})$$

or

$$I_{\downarrow o} = \epsilon\sigma T^4[a_o - b_o \exp(-2.3c_o e)]$$

where $I_{\downarrow o}$ is the counter radiation with clear skies; a_o, b_o, and c_o are empirical constants; T is the air temperature near the surface in °K; and e is the vapor pressure in millimeters of mercury (mm Hg) or millibars (1 mm Hg = 1.333 mb).

Since

$$I_o = I_\uparrow - I_{\downarrow o} \quad \text{and} \quad I_\uparrow = \epsilon\sigma T^4 ,$$

then

$$I_o = \epsilon\sigma T^4(1 - a_o + b_o 10^{-c_o e}) , \tag{4.5}$$

The values given for the constants differ from one investigation to the next, depending on how they are obtained. Listed values of a_o range from 0.710 to 0.820; b_o from 0.148 to 0.326; and c_o from 0.041 to 0.094 when the vapor pressure is in millibars. Geiger (1961) uses $a_o = 0.820$, $b_o = 0.250$, and $c_o = 0.094$.

The effect of the second term in equation (4.5) is small for large vapor pressures, exceeding 17 mb. Under these conditions, which are common in summer over most of the eastern half of the United States and in southern Arizona and California, the effective outgoing radiation with clear skies is less than 20 percent of the total infrared radiation sent out from the earth's surface. At the other extreme, when the air is very dry, the ratio $I_o/I_\uparrow$ rarely exceeds 0.30. The climatological average values for the United States are 0.226 in January and 0.182 in July.

2. *The Brunt Equation.* Brunt (1932) found that the simple relation

$$I_{\downarrow o} = \epsilon\sigma T^4(a + b\sqrt{e})$$

or

$$I_o = \epsilon\sigma T^4(1 - a - b\sqrt{e}) \tag{4.6}$$

fits his observations much better than the Ångström equation. He obtained $a = 0.256$ and $b = 0.065$. Other investigators have gotten values ranging from 0.34 to 0.71 for a and from 0.023 to 0.110 for b. The medians of twenty-two evaluations are $a = 0.605$ and $b = 0.048$, which are very close to the values used by Budyko (1956), $a = 0.61$ and $b = 0.050$.

3. *The Relative Humidity Equation.* J. E. McDonald (unpublished) computed the infrared flux using the Elsasser chart and mean monthly soundings (U.S. Weather Bureau, 1957) for January and July at fifty-three stations; forty-five in the United States, three in the Caribbean, and five in Mexico. His data are reasonably well cor-

related with the surface relative humidity *RH* in percent, with the following result for the two months combined.

$$I_o = \epsilon\,(0.165 - 0.000769\,RH)\ \text{ly min}^{-1}\,. \tag{4.7}$$

The standard error of estimate for this equation is 0.0085 ϵ ly min^{-1}.

4. *The Swinbank Equations.* Swinbank (1963), on the basis of measurements made in Australia and in the Indian Ocean, concludes that the counter radiation can be estimated to a high degree of accuracy from the surface temperature alone. He gives two relationships that can be expressed in the forms

$$I_o = \epsilon(0.245 - 0.214\sigma T^4)\ \text{ly min}^{-1} \tag{4.8}$$

and

$$I_o = \epsilon\sigma T^4(1 - 9.35 \times 10^{-6} T^2)\,. \tag{4.9}$$

5. *Other Equations.*

Robitzsch (1926): $I_o/I_\uparrow = 1 - (c_1 p + d_1 e)\,T^{-1}$

Elsasser (1942) and Loennquist (1954): $I_o/I_\uparrow = c_2 - d_2 \log e$

Anderson (1954): $I_o/I_\uparrow = c_3 - c_4 e + d_3 10^{d_4 e}$

Anderson (1954) and Budyko (1963*a*): $I_o/I_\uparrow = c_5 - d_5 e\,.$

In the Robitzsch equation, p is the air pressure.

Most of these equations give average daily values of I_o to an accuracy of 10 percent or better as long as the surface vapor pressure falls between 9 and 27 mb. For smaller or larger values, the errors may be great. Over 70 percent of the stations used by McDonald (unpublished) have a surface vapor pressure of less than 9 mb in January, and 23 percent have values exceeding 27 mb in July.

6. *The Integrated Elsasser Chart.* One of the main criticisms of the empirical equations is that they implicitly assume that the effective outgoing radiation at any given point is determined chiefly by surface conditions. They are successful partly because most of the counter radiation originates in the lowest kilometer of the atmosphere. Nevertheless, better results should be obtained by using a measure of the total water vapor content in an atmospheric column rather than the surface vapor pressure or relative humidity as the moisture variable. For this reason, the values of I_o obtained by McDonald for fifty-three North American stations from the Elsasser chart were graphically correlated with the mean monthly precipitable water vapor values for January and July given by Reitan (1960*a*) for the same stations. The result is Figure 16. An emissivity of 1.00 is assumed. When values of I_o taken from the chart are compared with those obtained by McDonald, the standard error of estimate is 0.0046 ly min^{-1}, corresponding to a relative accuracy of about 5 percent. The same data, when

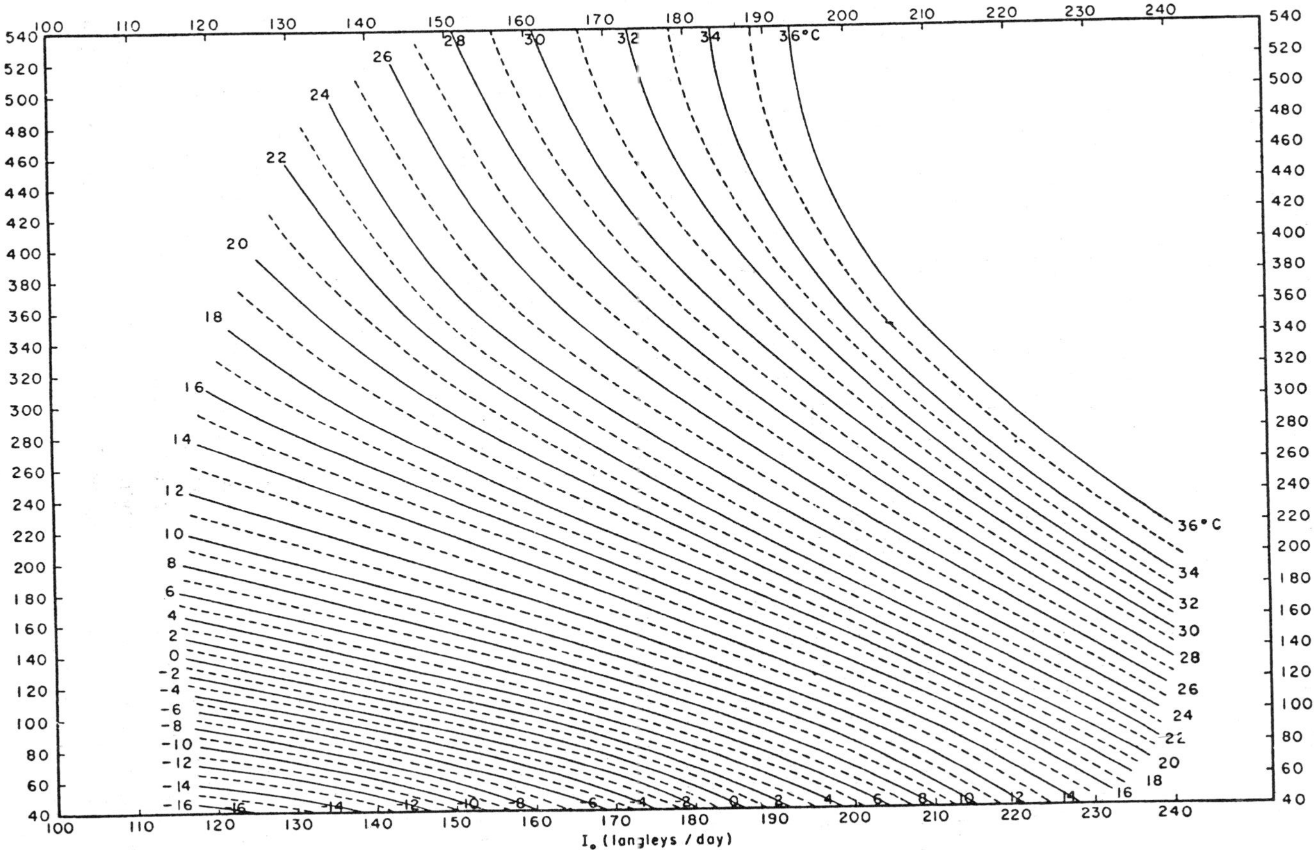

FIG. 16.—The integrated Elsasser chart. An emissivity of 1.0 is assumed

fitted to equations of the Brunt and Angström type, give standard errors of the order of 0.007 ly min^{-1}; so the chart does seem to be an improvement.

The over-all accuracy of the various charts and equations for estimating I_o can be judged by referring to data presented by Abraham (1960). He measured the atmospheric counter radiation on eighteen clear nights in 1958 at Tucson, Arizona, using a Suomi ventilated net radiometer and a Beckman-Whitely (B.W.) total hemispheric radiometer. To insure an unobstructed horizon, the roof of a tall building was used for the experiment. The values of $I\downarrow$ were compared with estimates made from the Elsasser and Kew radiation charts, using radiosonde observations taken within a half-hour of the radiometer readings at the Tucson International Airport, 7 miles to the south. The results, expressed as effective outgoing radiation; certain climatological data; and the average of the four measurements $\bar{I}_o$ are given in Table 11*a*. In general, the Beckman-Whitely radiometer and the Kew chart give average values about 20 percent higher than those obtained with the Suomi radiometer and the Elsasser chart. This is probably a good estimate of the present accuracy of measurements of the effective outgoing radiation with clear skies.

For comparison, climatological estimates of I_o derived from equations (4.5) to (4.9) and the integrated Elsasser chart (IEC) are given in Table 11*b*. The constants used in the Ångström and Brunt equations are $a_o = 0.82$, $b_o = 0.25$, $c_o = 0.094$, $a = 0.605$, and $b = 0.048$. An emissivity of 1.0 is assumed throughout. One measure of the relative accuracy of these values is the standard error of estimate s defined by

$$s = \left[\frac{1}{18}\sum_{1}^{18}(I_{oc} - \bar{I}_o)^2\right]^{1/2}$$

where I_{oc} is the computed value of I_o and $\bar{I}_o$ is the observed value, taken as the average of the four values obtained with the two radiometers and radiation charts and listed in the last column of Table 11*a*. The magnitude of s is given for each method in the last row of the two tables.

All climatological estimates, except those obtained from equations (4.8) and (4.9), compare favorably with the more direct measurements. The failure of Swinbank's two equations indicates that in the southwestern United States temperature alone is not a good predictor of I_o. It would be risky to conclude that the best estimates of I_o can be made using either the relative humidity equation or the integrated Elsasser chart, although Table 11*b* would seem to indicate otherwise. Other experiments of the same type, but in a different locality and perhaps with different values of the constants in the Brunt and Ångström equations, might lead to entirely different results. The main point here is that most empirical equations give reasonable estimates of I_o, even when applied to single observations.

TABLE 11*a*

Measured and Estimated Values of I_o for Tucson, Arizona

Date (1958)	T (° C)	e (mb)	RH (per-cent)	w_a (cm)	$I\uparrow$ (ly min⁻¹)	I_o (ly min⁻¹) Suomi	B.W.	Elsasser	Kew	$\bar{I}_o$
8/14.....	22.2	23.7	85.2	3.84	0.619	0.068	0.075	0.089	0.109	0.085
8/15.....	21.2	21.7	86.2	3.91	.610	.063	.086	.097	.120	.092
8/18.....	22.8	22.3	80.1	3.67	.623	.076	.106	.102	.125	.102
9/20.....	26.5	8.0	23.2	1.10	.655	.112	.128	.133	.148	.130
9/30.....	22.2	9.5	35.4	1.63	.619	.124	.161	.126	.140	.138
10/2......	24.5	11.9	38.6	1.47	.638	.157	.192	.145	.153	.162
10/7......	20.8	19.6	79.9	2.32	.607	.113	.141	.126	.132	.128
10/15.....	21.2	9.7	38.4	1.06	.610	.134	.170	.136	.144	.146
10/21.....	16.5	13.6	72.3	2.05	.571	.100	.124	.106	.123	.113
11/2......	11.8	5.5	39.6	0.60	.535	.132	.147	.129	.162	.142
11/3......	12.2	4.9	35.0	0.86	.538	.129	.147	.129	.155	.140
11/4......	13.2	5.2	34.3	0.54	.546	.136	.152	.126	.150	.141
11/5......	13.8	4.8	30.6	0.84	.551	.136	.169	.121	.142	.142
11/6......	14.8	5.9	35.0	1.01	.558	.130	.183	.128	.149	.148
11/12.....	11.5	8.3	60.8	1.23	.533	.118	.163	.121	.137	.135
12/12.....	10.2	5.5	44.0	1.10	.524	.111	.147	.109	.134	.125
12/13.....	10.2	5.9	47.4	0.89	.524	.117	.151	.124	.153	.136
12/18.....	7.2	3.6	35.1	0.024	0.502	0.118	0.146	0.120	0.161	0.136
Average.	....	......	....	....	....	0.115	0.144	0.120	0.141	0.130
s.......	....	......	....	....	....	0.016	0.019	0.013	0.015	......

TABLE 11*b*

Climatological Estimates of I_o for Tucson, Arizona

Date (1958)	I_o (ly min⁻¹) Ångström (4.5)	Brunt (4.6)	RH (4.7)	Swinbank (4.8)	Swinbank (4.9)	IEC
8/14.......	0.113	0.104	0.099	0.113	0.114	0.097
8/15.......	.111	.105	.099	.114	.116	.093
8/18.......	.113	.106	.103	.112	.113	.101
9/20.......	.147	.170	.147	.105	.105	.168
9/30.......	.131	.154	.138	.113	.114	.145
10/2........	.127	.147	.135	.108	.109	.155
10/7........	.111	.112	.104	.115	.117	.123
10/15.......	.128	.150	.135	.114	.116	.158
10/21.......	.110	.125	.109	.123	.123	.117
11/2........	.137	.152	.135	.131	.129	.149
11/3........	.143	.156	.138	.130	.128	.143
11/4........	.142	.157	.139	.128	.127	.154
11/5........	.148	.160	.141	.127	.127	.147
11/6........	.139	.156	.138	.126	.125	.145
11/12.......	.118	.138	.118	.131	.129	.129
12/12.......	.134	.149	.131	.133	.131	.130
12/13.......	.131	.146	.129	.133	.131	.137
12/18.......	0.148	0.153	0.138	0.138	0.133	0.151
Average...	0.129	0.141	0.126	0.122	0.122	0.136
s.........	0.015	0.017	0.012	0.021	0.021	0.012

The distribution of the effective outgoing radiation with clear skies over the United States is shown in Figures 17 and 18 for January and July, respectively. In January the highest values, exceeding 180 ly day^{-1}, are found in New Mexico, which is favored by relatively warm surface temperatures and low atmospheric water vapor content. The trough of low values in the central United States is associated with very low surface temperatures (reduced outgoing radiation) in its northern part and with high air humidity (increased counter radiation) in its southern part. In July the effective outgoing radiation with clear skies increases rapidly from the Pacific Coast inland, marking the transition from the cool, moist coastal regions to the warm, dry desert interior. The highest values are found in Nevada and western Oregon on the leeward side of the Sierra Nevada mountain range. In general, I_o is about the same in July and January in the eastern United States and considerably higher in July than in January in the western United States.

Equations (4.5) to (4.9) must be modified if the sky is not clear. As mentioned in the discussion of the Elsasser chart, clouds increase the counter radiation and, hence, decrease the effective outgoing radiation. The relationship between I_o, n, and I is usually expressed in the form

$$I = I_o(1 - kn^m) \tag{4.10}$$

although Geiger (1961) prefers

$$I_{\downarrow} = I_{\downarrow o}(1 + k_1 n^2)$$

where k, k_1, and m are constants to be determined from observations.

The quantity $(1 - k)$ represents the ratio of the effective outgoing radiation with overcast skies ($n = 1.0$) to that with clear skies. It is definitely a function of cloud height and cloud type, since the higher and thinner the clouds the less they will contribute to the counter radiation reaching the ground. The following values of $1 - k$ were obtained from data presented by Budyko (1956), Geiger (1959), and McDonald (unpublished).

Cloud Type	Height (m)	$1-k$
Cirrus	12,200	0.84
Cirrostratus	8,390	.68
Altocumulus	3,660	.34
Altostratus	2,140	.20
Stratocumulus	1,220	.12
Stratus	460	.04
Nimbostratus	92	.01
Fog	0	0.00

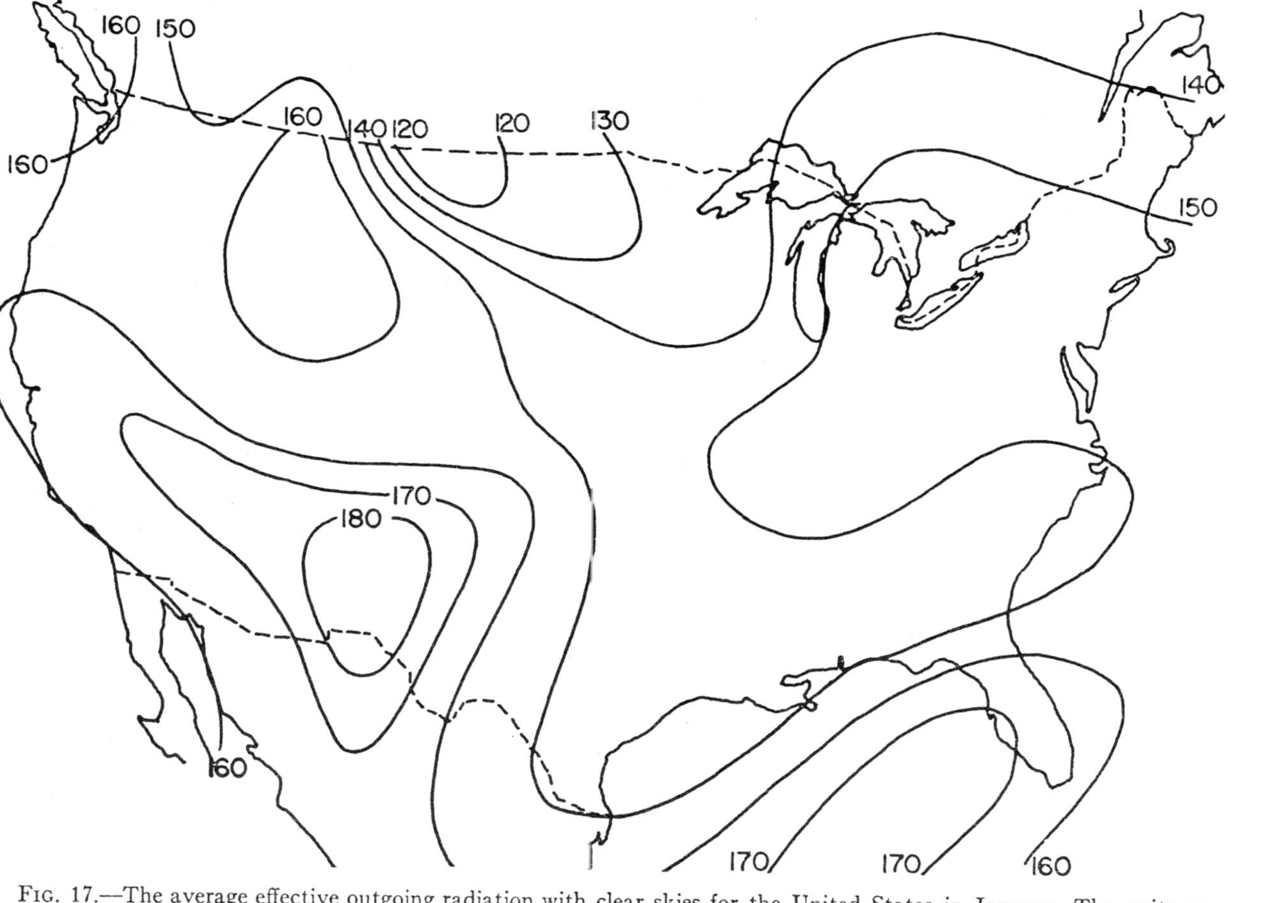

Fig. 17.—The average effective outgoing radiation with clear skies for the United States in January. The units are langleys per day. An emissivity of 1.0 is assumed.

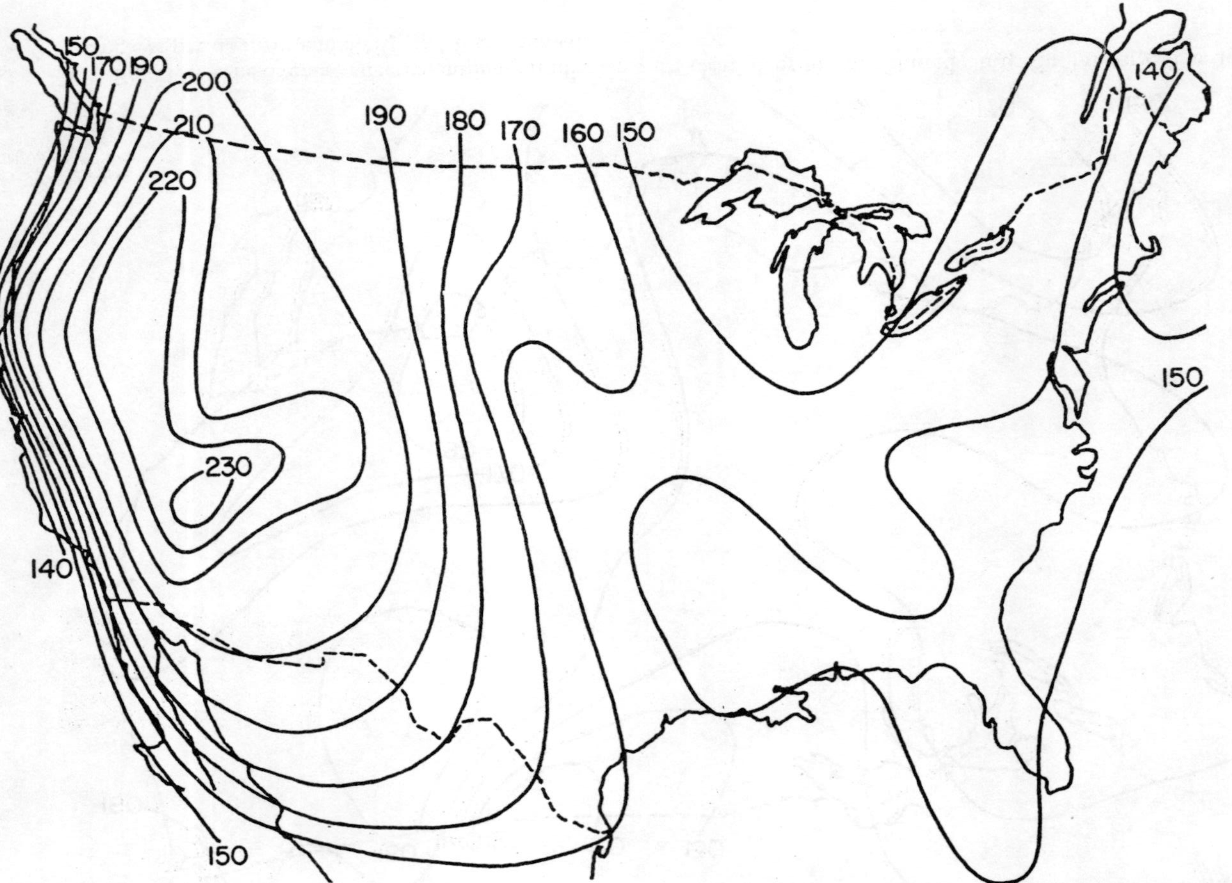

Fig. 18.—The average effective outgoing radiation with clear skies for the United States in July. The units are langleys per day. An emissivity of 1.0 is assumed.

The exponent m is usually set equal to 1.0. Budyko (1956) and Geiger (1961), however, both present data which indicate that it should lie between 1.5 and 2.7, with an average value close to 2.0. Nevertheless, Budyko (1963) still uses $m = 1.0$.

With $m = 2.0$, the ratio k_1/k equals $I_o/I_{\downarrow o}$, which averages close to 0.25. Hence, k_1 is about one-fourth of k. This, in turn, implies that the ratio $I_{\downarrow}/I_{\downarrow o}$ varies from 1.04 with a very high overcast to 1.25 with a thick fog. Cloud layers, then, as a rule do not increase the counter radiation by more than 25 percent over the clear sky value, if the atmospheric water vapor content is the same in both cases.

In computing the effective outgoing radiation or the net radiative flux from the earth's surface by any of the methods outlined in this section, it is customary to use the air temperature and vapor pressure at screen height (1 or 2 m) in the calculations rather than the true surface values. If these differ considerably, large errors may arise. Appendix 3 shows that the true effective outgoing radiation from the earth's surface I_s is related to that estimated from radiation charts or equations using temperature and humidity data at screen height I by the approximate relationship

$$I_s = I + 4\,\epsilon\sigma T^3(T_s - T) \tag{4.11}$$

where T_s is the true surface temperature. As indicated in the following table, the correction term can have the same magnitude as I when $T_s - T$ is greater than 10°C.

$T_s - T$ (° C)	$4\,\sigma T^3\,(T_s - T)$ (ly min^{-1})	$\Delta T_I/\Delta t$ (° C hour^{-1})
0	0	0
2	.018	3.5
4	.035	7.0
6	.053	10.5
8	.070	14.1
10	.088	17.6
15	.132	26.4
20	0.176	35.2

Appendix 3 also shows that the quantity $4\,\sigma\,T^3\,(T_s - T)$ is proportional to the net radiative flux into the air layer between the surface and screen height, the constant of proportionality being approximately 0.08. This, in turn, is related to the radiative heating or cooling of the layer through equation (3.11), which becomes

$$\frac{\Delta \bar{T}_I}{\Delta t} \simeq \frac{0.32\,\sigma T^3 (T_s - T)}{\rho\, c_p \Delta z} \tag{4.12}$$

where c_p is the specific heat of air at constant pressure. Values of $\Delta \bar{T}_I/\Delta t$, which represent the mean rate of temperature change within the air layer caused by the absorption or emission of infrared radiation, are listed in the preceding table for

$\Delta z = 100$ cm. The temperature changes are, for the most part, much larger than those normally observed. Almost as fast as the layer is warmed or cooled radiatively, energy is taken away or added by convection. According to equation (4.12), radiative cooling must prevail at night, when T_s is normally less than T, and radiative heating during the day, when T_s is greater than T.

The combination of equations (4.10) and (4.11) gives for the best estimate of the effective outgoing radiation from the earth's surface

$$I_s = I_o(1 - kn^m) + 4\epsilon\sigma T^3(T_s - T) \tag{4.13}$$

where I_o can be determined by any of the methods mentioned earlier. A practical method for using this equation will be described in chapter 8 and Appendix 6.

To this point, methods for estimating the total effective outgoing radiation from a plane surface with an unobstructed horizon have been discussed. In many cases it is important to know how this radiation is related to the outgoing radiation from a valley or canyon or other nonlevel terrain. Dubois (1929), Linke (1931), Lauscher (1934), and Hinzpeter (1957), among others, have discussed this problem in detail. Their work is summarized by Geiger (1959, 1961).

Because the effective radiating temperature of the atmosphere is usually lowest directly overhead, the net outgoing radiation is greatest in this direction and decreases with increasing zenith distance to zero on the horizon. Hence, a canopy or a large tree provides effective frost protection for the underlying surface by shielding it from the coldest portion of the sky toward which radiative losses are greatest. In early winter it is not unusual in some areas for grass to stay green and for flowers to remain in bloom under a tree long after the surrounding, exposed vegetation has died.

The variation of the net outgoing radiation with the zenith angle is shown below for data obtained by Brooks (1959) on three clear winter nights at Blue Hill, Massa-

Z (deg)	Brooks's Observation	Equation (4.14)
0	1.000	1.000
10	0.996	0.995
20	0.991	0.982
30	0.978	0.958
40	0.949	0.923
50	0.913	0.876
60	0.859	0.812
65	0.815	0.772
70	0.750	0.725
75	0.656	0.667
80	0.551	0.591
85	0.398	0.481
90	0.000	0.000

chusetts. The values are expressed in terms of the radiation toward the zenith. From a series of measurements by Dubois (1929), Linke (1931) found that the variation can be expressed quite adequately by

$$I_o(Z) = I_o(0) \cos^\gamma Z \tag{4.14}$$

where $I_o(Z)$ is the outgoing radiation in the direction Z, $I_o(0)$ is the outgoing radiation toward the zenith, and γ is a constant which depends on the surface vapor pressure e. As shown in the preceding table, equation (4.14) fits the Blue Hill observations fairly well if $\gamma = 0.3$. For the surface vapor pressure (3.97 mb) observed, this value is somewhat larger than what would be obtained from the relation suggested by Linke; that is,

$$\gamma = 0.11 + 0.0255\, e$$

where e is in millibars.

Süssenberger (1935) and Hinzpeter (1957) have suggested empirical relationships between $I_o(Z)$ and $I_o(0)$ that seem to fit the observations better than equation (4.14). This equation, nevertheless, is useful for studying the dependence of I_o on terrain features. Integrating it over the whole hemisphere gives

$$I_o = \frac{2\pi}{\gamma + 2} I_o(0),$$

which for $\gamma = 0$ reduces to the usual relationship between the flux I_o and the intensity of emission $I_o(0)$. If the sky is partially obscured as it would be at the bottom of a deep basin or in a circular valley, then the integration is only over the visible portion of the sky; that is,

$$I_{ob} = \frac{2\pi}{\gamma + 2}(1 - \cos^{\gamma+2} Z_1)\, I_o(0)$$

where Z_1 is the zenith distance to the rim of the basin. The ratio of the effective outgoing radiation from a basin I_{ob} to that from a flat plain is then

$$I_{ob}/I_o = (1 - \cos^{\gamma+2} Z_1)\,.$$

Values of this ratio are listed below for different values of Z_1 and for γ equal to 0.3 and 0.8. The same figures would apply to a tree obscuring the sky from the zenith to $90° - Z_1$. The effective outgoing radiation from a basin whose rim extends 40° above the horizon ($Z_1 = 50°$) will be only 64 to 71 percent of that from a flat plain. The ratio will be higher the greater the moisture content of the air.

These results help explain the unusual warmth experienced in deep valleys and canyons on hot summer days. For example, the average maximum temperature at the bottom of the Grand Canyon (elevation, 772 m) in July is 41.4°C, which is 12°C higher than that on the south rim of the canyon (elevation, 2,115 m) and almost the

Z_1 (degrees)	$\gamma=0.3$	$\gamma=0.8$
0..........	0.000	0.000
10..........	0.035	0.042
20..........	0.133	0.160
30..........	0.282	0.332
40..........	0.458	0.526
50..........	0.638	0.710
60..........	0.797	0.856
65..........	0.862	0.910
70..........	0.915	0.950
75..........	0.955	0.977
80..........	0.982	0.993
85..........	0.996	0.999
90..........	1.000	1.000

same as that at Yuma, Arizona (elevation, 52 m). Practically all infrared radiation from the strongly heated walls is retained and used to warm the air. At night, energy radiated and conducted from valley slopes will provide some frost protection. This will, however, often be over-balanced by cold air drainage from higher elevations.

Lauscher (1934) has made similar calculations for a wide range of terrain types, including elongated valleys, cliffs, and streets.

5/ THE RADIATION BALANCE

Averaged over the globe, the earth's surface absorbs about 124 kly of solar radiation each year and, in turn, effectively radiates 52 kly of longwave energy to the atmosphere. The difference between these two figures, 72 kly, is the net radiative balance R of the earth's surface. It is usually called the radiation balance or the net radiation and is symbolically given by

$$R = (Q + q)(1 - a) - I \tag{5.1}$$

where $Q + q$ is the sum of the direct and diffuse solar radiation incident on the earth's surface, a is the surface albedo, and I is the effective outgoing radiation from the surface.

The radiation balance R_a of the atmosphere may be defined similarly. Since the atmosphere absorbs only 45 kly of solar energy per year and radiates 117 kly of longwave energy, it has a negative radiation balance of -72 kly year^{-1}.

Thus the atmosphere loses just as much radiative energy in a year as the earth's surface gains. The radiation balance of the whole system, surface plus atmosphere, R_g, is therefore zero. This must be so; otherwise, the earth as a planet would be getting warmer or cooler.

Although the global radiation balance is zero averaged over the year, it will generally not equal zero either seasonally or annually in a given latitude zone. The annual latitudinal distributions of R, R_a, and R_g are shown in the top part of Figure 19.

The atmosphere is uniformly a radiative heat sink at all latitudes, while the earth's surface, except near the poles, is a heat source. Energy must therefore be transferred from the surface to the atmosphere to keep the surface from warming (at a rate of about 250°C per day in the upper centimeter) and the atmosphere from cooling (at a rate of about 1°C per day). As shown in chapter 8, this vertical heat exchange occurs mainly by evaporation of water from the surface (heat loss) and condensation in the

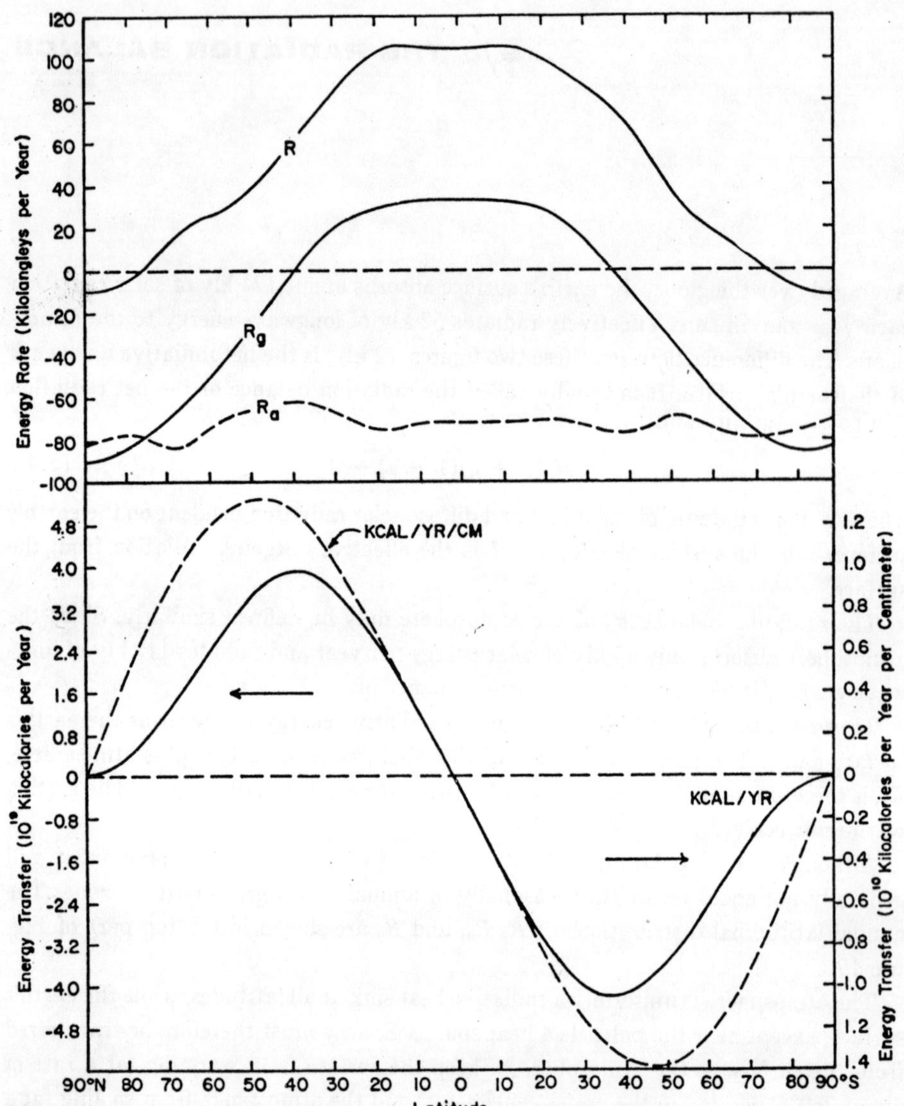

FIG. 19.—The average annual latitudinal distribution of the radiation balances of the earth's surface R, of the atmosphere R_a, and of the earth-atmosphere system R_g, in kilolangleys per year (*top*) and the poleward energy flux in kilocalories per year and in kilocalories per year per centimeter of latitude circle (*bottom*).

atmosphere (heat gain) and by conduction of sensible heat from the surface and turbulent diffusion into the atmosphere (convection).

Poleward of 40° the radiative deficit of the atmosphere exceeds the surplus of the surface, which leads to a negative radiation balance for these parts of the globe. Conversely, equatorward of 40° the radiative surplus of the surface exceeds the deficit of the atmosphere and results in a positive radiation balance. Poleward of 40° the sink is stronger than the source; equatorward of 40° the source is stronger then the sink.

In order to keep the poles from getting colder and the tropics from getting warmer, energy, therefore, must be transferred meridionally from lower to higher latitudes.

TABLE 12

ESTIMATES OF THE ANNUAL POLEWARD HEAT FLUX IN THE NORTHERN HEMISPHERE

(kcal year^{-1} $\times$ 10^{-19})

Latitude	Simpson (1929)	Baur and Philipps (1935)	Gabites (1950)	Raethjen (1950)	Houghton (1954)	Budyko (1956)	Berlyand (1956)	London (1957)	Figure 19
90° N.	0.00	. . . .	0.00	. . . .	0.00	. . . .	. . . .	0.00	0.00
80.	0.35	. . . .	0.33	. . . .	0.34	. . . .	. . . .	0.29	0.35
70.	1.34	. . . .	1.15	. . . .	1.24	. . . .	0.91	1.08	1.25
60.	2.36	1.89	2.11	3.07	2.44	2.35	2.13	2.09	2.40
50.	3.40	2.79	2.89	. . . .	3.51	3.31	3.25	2.94	3.40
40.	3.96	3.34	3.29	5.01	4.06	3.89	3.91	3.43	3.91
30.	3.71	3.21	2.84	. . . .	3.82	3.52	3.98	3.16	3.56
20.	2.81	2.42	1.83	3.94	2.80	2.74	3.39	2.44	2.54
10.	1.57	1.37	0.62	. . . .	1.48	1.50	2.05	1.21	1.21
0.	0.18	0.00[1]	−0.55	0.00[1]	0.00[1]	0.00[1]	0.00[1]	−0.09	−0.26

[1] Assumed values.

This horizontal heat exchange, which is induced partly by differential heating of continents and oceans, is carried out mainly by the poleward transfer of sensible heat by the atmospheric circulation and ocean currents and by the release of latent heat through condensation of water vapor carried poleward in the atmosphere. The importance of each mechanism will be discussed in chapter 8.

For the present, we conclude that a two-way heat transfer must exist—from the surface to the atmosphere and from the equator to the poles. The transfer must occur in such a manner that no part of the earth warms or cools noticeably over a period of a year. The magnitude of the meridional heat flux required to preserve this balance, expressed in both kilocalories per year and kilocalories per year per centimeter of latitude circle, is shown in the bottom part of Figure 19. Positive values indicate a northward flux; negative values a southward flux.

The meridional flux is greatest across 40° to 50° and is slightly larger in the southern hemisphere than in the northern hemisphere. The maximum magnitude of 4.2 × 10^{19} kcal year^{-1} across 40°S, however, is equal to less than 5 percent of the solar radiation absorbed by the earth-atmosphere system in one year (85.7 × 10^{19} kcal).

There is still considerable doubt as to the exact magnitude of the meridional heat flux, especially in the southern hemisphere. Satellite observations will help resolve this problem, but probably not for several years. Estimates for the northern hemisphere given by several authors are summarized in Table 12 and compared with the values from Figure 19. The latter are based mainly on the data of Houghton (1954) and Budyko (1956).

It is interesting to note how well the latest figures agree with those obtained by Simpson (1929) forty years ago. The estimates of Gabites (1950) and Raethjen (1950) appear to be out of line when compared with the results of the other investigators. Nevertheless, our understanding of the general circulation of the atmosphere and oceans is still too incomplete to decide who is right and who is wrong. Significant changes will occur in the results if the usual assumption of no net energy flux across the equator proves to be incorrect.

6/ RADIATION INSTRUMENTS

Many instruments are available for measuring the components of the radiation balance. These can be grouped according to whether they measure the direct-beam solar radiation at normal incidence (pyrheliometers), the solar radiation received from the whole hemisphere (pyranometers), infrared radiation (pyrgeometers), both infrared and solar radiation from a single hemisphere (pyrradiometers), or the radiation balance (net radiometers). The nomenclature is that suggested by the World Meteorological Organization in 1963 but is not universal. For further information the reader is referred to publications by the British Meteorological Office (1956), the Special Committee for the International Geophysical Year (1958), Gates (1962), and the World Meteorological Organization (1963).

PYRHELIOMETERS

Pyrheliometers are instruments designed to measure the direct-beam solar radiation at normal incidence, usually called the solar intensity. The blackened receiving surface is oriented perpendicular to the solar beam and is either inserted in a blackened tube or surrounded by a series of spaced diaphragms arranged so only radiation from the sun and a narrow annulus of sky is intercepted. Observations are made only when the sky is completely clear to a distance of at least 20° from the sun.

Pyrheliometers are the most accurate of all radiation instruments and as such they are commonly used as calibration standards. They are also expensive and usually found only at special research laboratories and observing stations. Pyrheliometric readings are taken regularly in the continental United States only at Albuquerque, New Mexico; Blue Hill, Massachusetts; Madison, Wisconsin; Omaha, Nebraska; and Tucson, Arizona. The data are used mainly in studies of atmospheric turbidity.

Abbot silver disk pyrheliometer.—This instrument maintains its calibration for many years and is used in the United States as a standard pyrheliometer for the calibration of secondary, routine pyrheliometers and pyranometers. It has a blackened silver disk supported by fine steel wires at the lower end of an extended copper tube. The bulb of a very accurate mercury thermometer is inserted into the disk (Fig. 20).

The solar intensity is measured with the aid of equation (3.11), written in the form

$$Q_n = Cd\,\frac{\Delta \bar{T}}{\Delta t}$$

where C is the heat capacity of the disk and d is its thickness. The rate of temperature change is determined by using a triple shutter to expose the disk to and shade it from

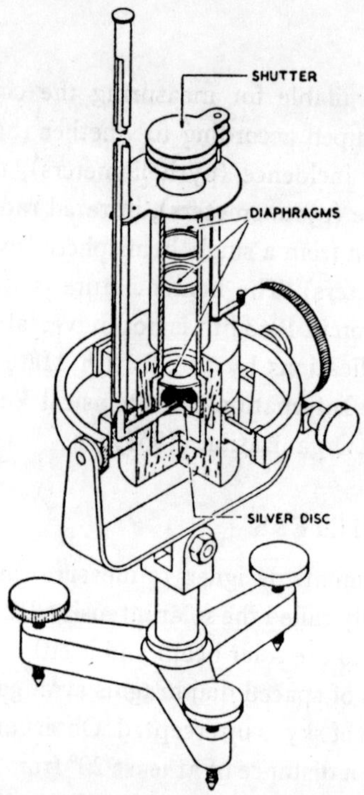

Fig. 20.—The Abbot silver disk pyrheliometer. From *Handbook of Meteorological Instruments*, Part I. (British Meteorological Office, 1956.)

the solar beam alternately at regular 2-min intervals. The timing must be very precise; a consistent error of 1 second may result in an error of 1 percent in the final result. Corrections must be applied to reduce each reading to a standard air temperature of 20°C and a standard disk temperature of 30°C.

The calibration constant Cd of the silver disk pyrheliometer is determined by comparing readings of the instrument with those of a water-flow calorimetric primary standard pyrheliometer housed at the Smithsonian Institution in Washington, D.C. Under ideal conditions the silver disk pyrheliometer has an accuracy of better than ±0.5 percent.

Ångström compensation pyrheliometer.—The Ångström pyrheliometer is one of the best known and most reliable instruments available for measuring the solar intensity. It is widely used in Europe as a standard instrument for calibrating secondary pyrheliometers and pyranometers and has an accuracy of the order of ±0.5 percent.

This instrument has a shaded, thin, blackened manganin strip that is heated electrically to the same temperature as a similar strip exposed to solar radiation (Fig. 21). Under steady-state conditions, that is, when both strips are at the same

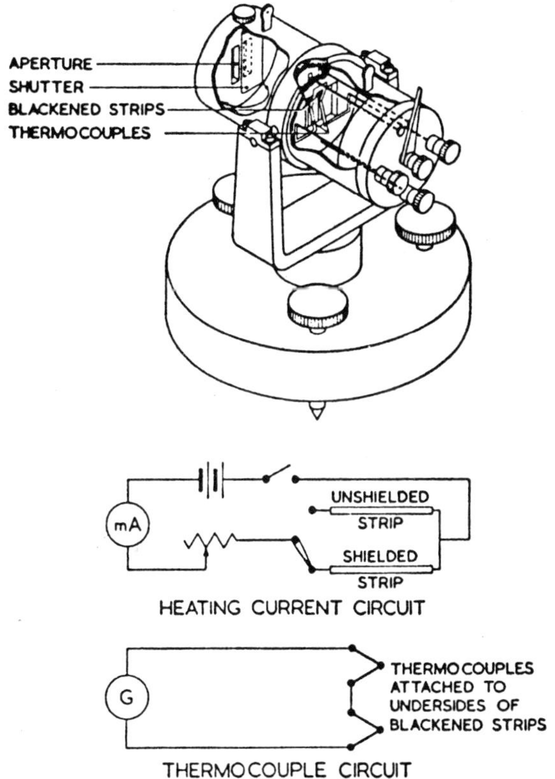

FIG. 21.—The Ångström compensation pyrheliometer. From *Handbook of Meteorological Instruments*, Part I. (British Meteorological Office, 1956.)

temperature, the rate at which electrical energy is used to heat the shaded strip must equal the rate at which solar energy heats the exposed strip. Hence,

$$Q_n = ki^2$$

where i is the heating current and k is a dimensional and instrument constant that depends on the resistance, width, and absorptivity of the manganin strips. It is possible to determine k directly; however, in practice most Ångström pyrheliometers are calibrated by comparison with a standard instrument housed in Stockholm.

Other types.—There are a number of other pyrheliometers, most of which are commonly used for routine observations. These are not as accurate as either the Abbot or the Ångström instruments and must be recalibrated periodically. They are the Eppley (normal incidence) pyrheliometer, which is used at pyrheliometric stations in the United States; the Michelson bimetallic pyrheliometer (actinometer), which is a portable self-contained instrument and is especially suitable as a traveling sub-standard, if great care is exercised in its transportation; and the Linke-Feussner pyrheliometer (actinometer), which can also be used to measure thermal radiation from selected parts of the sky at night and from the ground.

PYRANOMETERS

Pyranometers are used much more than are pyrheliometers. They measure the total shortwave radiation from the sun and sky incident on a horizontal surface at the ground ($Q + q$). Unlike pyrheliometers, pyranometers are exposed continuously and in all kinds of weather. Therefore, they must be sturdy and mounted securely. The receiver is inclosed in a glass or quartz casing that must be kept clean and dry. To provide the best possible record, a pyranometer should be located so that a shadow will not be cast on it at any time. It should not be near light-colored walls or other objects likely to reflect sunlight onto the receiver, and it should not be exposed to artificial radiation sources.

The glass or quartz dome used with pyranometers not only protects the receiver from wind and rain but also transmits only shortwave radiation, between 0.35 and 2.8μ for colorless glass and between 0.25 and 4.0μ for quartz. The receiver surface usually has at least two sensing elements, one blackened to absorb a large proportion of the incident radiation and the other coated with a white paint of high shortwave reflectivity. Appendix 4 shows that the temperature difference between the two elements is proportional to the incident solar radiation, with the constant of proportionality being weakly dependent on the temperature of the sensor.

There are about seventy-five pyranometric stations in the continental United

States. Most have been in continuous operation since 1952. The network is dense enough to permit fairly detailed mean monthly maps of the solar radiation reaching the ground, such as in Figures 9 and 10.

Eppley pyranometer (180° pyrheliometer).—This is the standard instrument for solar radiation measurements in the United States. The sensor consists of two thin, flat, concentric silver rings. The inner one is coated black and the outer one is coated white (magnesium oxide). (See Fig. 22.) The temperature difference between the two

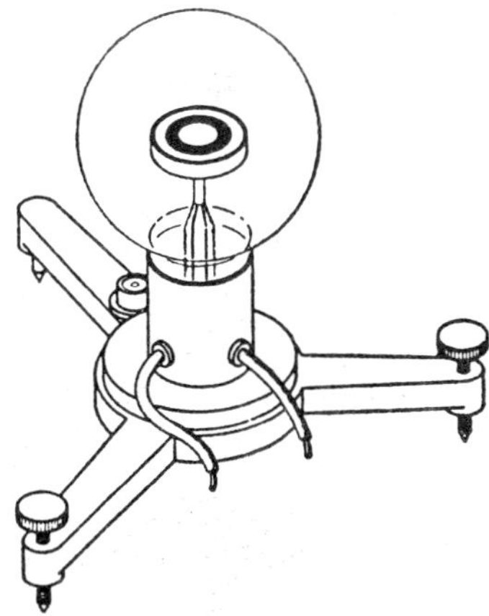

FIG. 22.—The Eppley pyranometer (180° pyrheliometer). From *Handbook of Meteorological Instruments*, Part I. (British Meteorological Office, 1956.)

rings is measured with a large number of thermojunctions (16 or 50, depending on the accuracy required), which are in good thermal contact with the lower surfaces of the rings. The receiving assembly is inclosed in a specially blown spherical glass bulb that is filled with dry air and that limits the radiation response to wavelengths shorter than 3.5μ.

Moll-Gorcznski pyranometer (solarimeter).—The Moll-Gorcznski pyranometer is commonly used in Europe and is considered somewhat more accurate than the Eppley pyranometer. The blackened rectangular sensor has alternate thin strips of manganin and constantan, forming a series of fourteen thermojunctions (see Fig. 23). Instead of being coated white, one set of junctions is in good thermal contact with a heavy brass block set nearly flush with the sensor surface; as a result, its temperature changes only slowly. The assembly is inclosed in two concentric hemispherical glass domes, 2-mm thick and 30 and 50 mm in diameter. The main purpose of the double domes is to reduce convective heat losses. The Eppley Corporation has used this feature in a new temperature-compensated model of their instrument.

Robitzsch bimetallic pyranograph (*actinograph*).—This is a clock-driven instrument whose dome-covered receiving surface has three parallel bimetallic strips; the outer two are white-coated and the inner one is black-coated. The temperature difference between the strips is recorded through a mechanical linkage. This instrument has a long lag time, about 10 to 15 min for 98 percent response compared to 30 seconds for the electrical pyranometers. As a result, it is suitable only for obtaining daily totals

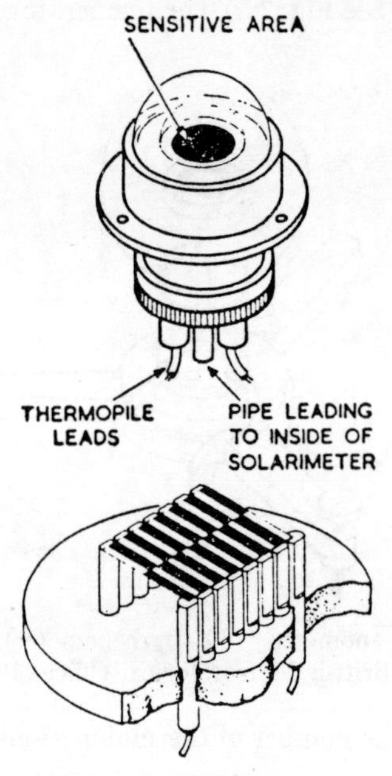

FIG. 23.—The Moll-Gorcznski pyranometer (solarimeter). From *Handbook of Meteorological Instruments*, Part I. (British Meteorological Office, 1956.)

of solar radiation and even these will be accurate to no more than 5 to 10 percent because of other difficulties inherent in the instrument. There are several models of the Robitzsch pyranograph, one of which is called a pyrheliograph.[1]

Bellani spherical pyranometer.—This instrument is not a pyranometer in the true sense of the word, because its receiver surface is spherical rather than flat. It can be useful, however, in estimating the total solar radiation incident on a freely exposed object such as a single plant or tree. As shown in Figure 24, the Bellani pyranometer has two concentric glass spheres mounted on the end of a burette graduated to 40 ml. The space between the spheres is evacuated in order to eliminate heat exchange by

[1] The pyrheliograph is available from Science Associates, Inc., Princeton, N.J.

convection. The inner sphere, which acts as the receiver surface, is covered with a black-coated copper shell. A reservoir of pure ethyl alcohol within the inner sphere is open to the burette through a small capillary tube. As the copper shell absorbs solar radiation and is heated, some of the liquid evaporates, only to condense again at the bottom of the burette. The amount of condensate in a given time is directly related to the total shortwave radiation intercepted by the copper shell. One model of the pyranometer is designed to be placed in a metal cylinder and sunk into the soil, so that the top of the inner sphere is just at ground level. In this case no reflected radiation

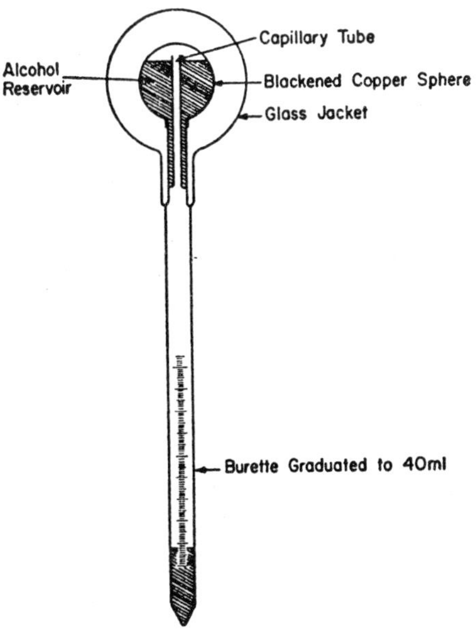

FIG. 24.—The Bellani spherical pyranometer

from the ground is received. Readings of the burette are taken at hourly or daily intervals. After each reading, the instrument is inverted to allow the alcohol to return to the reservoir. According to Courvoisier and Wierzejewski (1954), the Bellani pyranometer is capable of measuring total daily amounts of radiation to an accuracy of at least ±3 percent. Shaw and McComb (1959) obtained equally good results using a water-filled pyranometer.

Additional uses for pyranometers.—Besides measuring the total solar radiation incident on a horizontal surface, pyranometers can also measure the total solar radiation incident on an inclined plane, the diffuse sky radiation, the solar intensity, and the surface albedo.

To measure the diffuse sky radiation q, a shading ring or disk screens the receiver from the direct solar rays. The shading ring is mounted on an adjustable bar attached to the pyranometer and sloping upward toward the north (in the northern hemisphere) at an angle equal to the latitude of the station. The ring, whose plane is normal to the bar, can be moved up or down the bar as the solar declination changes. When properly set, the rim of the ring traces the path of the sun across the sky, thus shading the receiver from the direct solar rays from sunrise to sunset. Readings taken with this apparatus must be semi-empirically corrected to take into account diffuse radiation intercepted by the ring. Continuous readings of diffuse radiation with a shading disk, rather than a ring, can be obtained by attaching the disk to an equatorial mounting similar to that used with pyrheliometers. Whether a ring or disk is used, the entire glass bulb of the pyranometer should be shaded in order to prevent errors arising from multiple reflections at the glass surface.

The solar intensity Q_n can be obtained indirectly either by using two pyranometers, one shaded and the other unshaded, or by alternately shading and unshading a single pyranometer. In either case, readings of both $Q + q$ and q are obtained. Their difference is the direct-beam solar radiation on a horizontal surface Q, which when divided by the cosine of the solar zenith angle gives the solar intensity. This is the method used to check the calibration constant of pyranometers, since Q_n can be measured directly with a pyrheliometer.

The surface albedo is determined either with two pyranometers, one of which faces downward, or with a single pyranometer, which is periodically inverted. The upward-facing pyranometer records the total sunlight $Q + q$; the downward-facing instrument records only the reflected sunlight $(Q + q)\alpha$. The albedo is obtained by taking the ratio of the two readings. Measurements are made near the ground, if the local albedo is desired, or from an airplane or balloon, if the mean albedo of a large area or of cloud decks is required. In the latter case special precautions must be taken to make sure that the receiving surfaces remain horizontal.

PYRGEOMETERS, PYRRADIOMETERS, AND NET RADIOMETERS

As shown in Appendix 4, the temperature difference between two identical blackened sensing elements placed back to back in the free air a short distance above the ground, with one facing upward and the other facing downward, is directly proportional to the net radiation R. That is,

$$R = C(T_u - T_d) \tag{6.1}$$

where T_u and T_d are the temperatures of the upper and lower elements, respectively. If the surface of the lower element is covered with highly polished aluminum, which has a very low emissivity in the long wavelengths and a high albedo in the short wavelengths, it follows that

$$Q + q + I_{\downarrow} = C(T_u - T_d) + \sigma T_d^4 . \tag{6.2}$$

Equation (6.1) is the governing equation for net radiometers and equation (6.2) for pyrradiometers, which are sometimes called effective pyranometers. Obviously, radiometers can be used as pyrradiometers by shielding one surface of the receiving element from radiation exchange. Both instruments become pyrgeometers at night when there is no solar radiation.

The parameter C, which appears in equations (6.1) and (6.2), is not a constant but depends on the physical properties and temperature of the sensor and on the free-air wind speed, which determines the rate at which the sensor is warmed or cooled by conduction and convection. Convection is by far the most important variable and must either be eliminated or made constant if the instrument is to have any practical value. Several methods have been proposed for doing this, among which are the following:

1. A jet of air is blown equally over both faces of the freely exposed sensor, thus providing an artificial but constant convective heat loss which minimizes the effect of the variable natural convection. Natural convection, however, cannot be completely eliminated, even with a jet speed of 20 to 25 m sec^{-1}. As a result, the output of this instrument often varies rapidly and significantly, especially when a natural wind of more than 8 m sec^{-1} blows normal to the forced-air stream. The ventilation must be uniform over both faces of the sensor; otherwise, the response of the two sides to a given radiation flux will be unequal.

2. The sensor is inclosed in a sphere transparent to both shortwave and longwave radiation. Glass is obviously not suitable because of its high absorptivity in the infrared. Thin polyethylene is often used. A 0.1-mm film of this material transmits about 85 percent of the incident radiation between 0.3 and 100μ. The strongest absorption bands are in the infrared at 3.5, 6.9, and 14μ. As a result, shortwave radiation is more readily transmitted than longwave radiation, hereby producing an apparent increase in the calibration constant with increasing wavelength. Other plastics, including polystyrene, are sometimes used, but generally these are inferior to polyethylene.

3. Both faces of the sensor are kept at the same temperature by introducing a compensation current to warm the cooler of the two. If it is then assumed that the

convective heat loss is the same from both faces, the magnitude of this current will be directly proportional to the square root of the desired radiative flux.

Ångström compensation pyrgeometer.—This was one of the first infrared detectors built and is now mainly of historical interest. It can be used only at night. The sensor has four thin manganin strips, two blackened and two gold-plated. The blackened strips radiate freely to the night sky and cool; the gold-plated strips, which have a very low infrared emissivity, remain at air temperature. The cooling of the black strips is compensated by an electrical heating current that keeps all four strips at the same temperature. In this case, equation (6.2) takes the form

$$I_{\downarrow} = \epsilon_b \sigma T_b^4 - C_1 i^2$$

where ϵ_b and T_b are the emissivity and temperature, respectively, of the black strips; C_1 is a calibration constant; and i is the compensation current. Best results are obtained on dry nights with little or no wind.

Hofmann radiometer.—This is a compensation-type radiometer. The sensor is a single freely exposed blackened disk with heating coils and thermopiles or thermistors mounted on the inner sides of the two faces, which are kept at the same temperature. It is assumed that the convective heat loss from the upper surface is the same as that from the lower surface. In this case, the net radiation is directly proportional to the square of the heating current. Actually, with the same wind speed and thermal stratification, convective heat losses during the day will be somewhat greater from the upper face than from the lower face. The resulting error can be minimized by proper calibration of the instrument. For field use, instead of constantly adjusting the heating current to keep the two faces of a single sensor at the same temperature, it is more practical to use two sensors and apply the same fixed amount of heat to the upper face of one and the lower face of the other. In this case

$$R = \frac{T_u - T_d + T'_u - T'_d}{T_u - T_d - T'_u + T'_d}\, h$$

where the unprimed temperatures refer to the sensor whose upper face is heated and the primed temperatures to the sensor whose lower face is heated; h is the heating current expressed in energy units. It is assumed that the wind speed-dependent calibration constant in equation (6.1) is the same for both sensors at all times.

Suomi ventilated net radiometer.—The sensor of this instrument is a blackened glass microscope slide, 25 × 75 × 1.1 mm, wound with constantan wire, half of which is copper-coated. It is exposed without any protective covering and is subjected to a sustained blast of air from a small centrifugal blower. The air speed is of the order of 20 to 25 m sec^{-1}. The radiometer has a specially designed vane in the nozzle throat and an electric heater on the radiation plate. By using the latter to heat the sensor in a field of uniform radiation and adjusting the vane, one can accurately equalize the cooling power on each side of the plate. The instrument is relatively large and must be mounted at least 1.5 m above the ground in order to minimize the fraction of the surface viewed by the lower face of the sensor which is covered by its own shadow.

Other ventilated radiometers include the Courvoisier net radiometer, which is especially suitable for instantaneous measurements of both the upward- and downward-directed components of the radiation balance and is recommended as the standard instrument in Europe, and the Beckman and Whitley thermal radiometer, which was developed by Gier and Dunkle at the University of California and is widely distributed in the United States. The latter instrument is frequently used as a pyrradiometer or total hemispherical radiometer.

Funk radiometer.—This is a polyethylene-shielded radiometer. The hemispheres are 0.05-mm thick and kept inflated with dry nitrogen gas. They are waterproofed with a thin coating of silicone oil, which unfortunately reduces the transmission of the polyethylene film by introducing several weak absorption bands. The sensor is a blackened thermopile of 250 junctions, which has a response time of about 1 min and a wind dependence of less than 1 percent over a wind speed range from 0 to 15 m sec^{-1}. Thin white strips of magnesium oxide, which reflects shortwave radiation and absorbs longwave radiation, are applied to both faces of the sensor. This helps to stabilize the calibration constant. The instrument is quite small, about 5 cm in diameter, and can be used close to the ground. A miniaturized version, suitable for biological studies, has a diameter of only 1 cm.

Other radiometers using this same principle are the Schulze radiometer, which is really two pyrradiometers, one facing upward and one downward, each with a hemispherical shield of polyethylene; the Georgi universal radiometer, which gives instantaneous values for all components of the radiation balance by successive measurements and was developed especially for expeditions where continuous recording is not possible; and the Thornthwaite miniature net radiometer, which is similar to the Funk radiometer but is not as sensitive. An improved model of the latter has been developed by Fritschen at Tempe, Arizona.

Economical radiometer.—This is a do-it-yourself polyethylene-shielded radiometer. It is especially popular in the United States, where it is widely used in meteorological, agricultural, and hydrological research. A simplified cross-sectional view of the upper unit of an instrument discussed in detail by Swan, Federer, and Tanner (1961) is shown in Figure 25. The sensor in this case is a dial thermometer with a 15-cm spike stem, which is embedded between two sheets of copper (or aluminum). The sheets are in good thermal contact with a high quality glass plate and rest on fiber glass and stafoam insulation. The lower surface of the glass plate is sprayed with heavy-duty black plastic enamel to provide nearly total absorption of all incident radiation. Two sheets of 1-mil (0.0254 mm) clear polethylene film, 1 cm apart, are mounted above the plate on a plywood tray. Since polyethylene expands when exposed to a strong radiation source, such as the sun, it is necessary to support both films with crossed nylon threads. The unit is inclosed in a standard aluminum chassis available at radio supply stores. The complete instrument has two such units, one facing upward and one facing downward. They are usually separated by a distance of about 20 cm, so that with natural ventilation the entire outside boundary of each unit is at or near air temperature. All components of the radiation balance can be measured

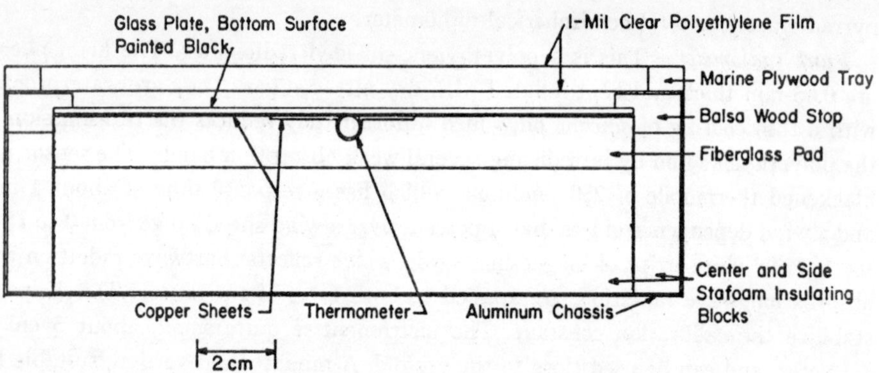

FIG. 25.—The economical net radiometer

with the economical radiometer if two instruments are used and the inner surfaces of the glass plates of one are silvered and painted white rather than blackened.

The economical radiometer is not as accurate as either the Funk or Suomi radiometers, partly because of the large lag time of the instrument and partly because of the use of flat, rather than hemispherical, ventilation shields. When corrections are made for these two effects, estimates of solar radiation and daytime net radiation will agree with readings made with the Eppley pyranometer and the Suomi net radiometer within 5 to 7 percent during clear and completely overcast periods and within 10 to 13 percent during partly cloudy periods, if the sun altitude is greater than 25°.

In spite of its drawbacks, the economical radiometer is a very useful and practical instrument. It is relatively easy to build, costs only a fraction of the price of the more precise radiometers, and is quite diversified in its possible applications. When thermistors or thermocouples are used instead of thermometers, the economical radiometer can be attached to a modified radiosonde unit for studying the vertical distribution of nighttime radiative cooling in the free atmosphere. Further information on the construction and applications of the economical radiometer can be obtained from papers by Suomi, Staley, and Kuhn (1960), Tanner, Businger, and Kuhn (1960), and Swan, Federer, and Tanner (1961).

CALIBRATION OF PYRGEOMETERS, PYRRADIOMETERS, AND NET RADIOMETERS

During the day, pyrradiometers and net radiometers can be calibrated with a pyranometer or a pyrheliometer. If a pyrheliometer is available and the sky is cloudless (or nearly so), the receiver of the radiometer is alternately shaded from and exposed to the sun for periods longer than the time of response of the instrument. The shading is done with a small plate held some distance away from the sensor. During this interval, the solar intensity Q_n is measured with the pyrheliometer. The difference between the radiometer readings when the receiver is shaded and unshaded is $Q_n \cos Z$.

If only a pyranometer is available, both it and the radiometer are exposed alternately to the total radiation and to all but the direct-beam radiation. Thus,

	Pyranometer	Pyrradiometer	Net Radiometer
unshaded	$Q+q$	$Q+q+I_{\downarrow}$	R
shaded	q	$q+I_{\downarrow}$	$R-Q$
unshaded-shaded	Q	Q	Q

At night a radiometer can be calibrated by shielding one face of the sensor from infrared radiation with a sheet of highly polished aluminum. The other face is exposed over a water surface of known temperature T_w. It then follows [eq. (6.2)], that

$$I_{\uparrow} = \epsilon_p \epsilon_w \sigma T_w^4 = C(T_d - T_u) + \sigma T_u^4$$

where ϵ_w is the emissivity of the water surface and ϵ_p is the emissivity of the downward-facing blackened plate for the wavelengths emitted by the water. Both emissivities are approximately equal to 0.98. The calibration constant C can be determined if, besides T_w, the temperature (or voltage) difference $T_d - T_u$ and the temperature T_u of the upward-facing shielded plate are known.

7/ THE WATER BALANCE AND THE HYDROLOGIC CYCLE

THE WATER BALANCE OF THE EARTH'S SURFACE

The water balance equation of the earth's surface is simply a mathematical formulation of the part of the hydrologic cycle that deals directly with the air-land or air-water interface. Consider a column of soil extending from the surface to that depth where the vertical moisture exchange is practically absent. The net rate g at which the moisture content of this column is changing is equal to the sum of the rates at which moisture is being added by precipitation r, by deposition of dew D, and by the horizontal flux of moisture into the column from the local environment f_i minus the rate at which moisture is being lost by evaporation E and by the horizontal flux of moisture out of the column f_o. Thus,

$$g = r + D + f_i - E - f_o . \tag{7.1}$$

Usually dew formation is negligible, with an upper limit probably in the vicinity of 1 mm per night. The difference $f_o - f_i$ is the net runoff Δf from the column and includes both surface and subsurface runoff. The evaporation component E includes transpiration from grass, plants, and trees and is sometimes called evapotranspiration. The equation also says that the precipitation falling on any given surface is either evaporated, stored in the soil column, or runs off,

$$r = E + g + \Delta f . \tag{7.2}$$

The water balance equation also applies to oceans, lakes, and reservoirs. Only here Δf represents the horizontal redistribution of water plus any runoff from surrounding streams or rivers and g represents the change in water level.

Over a period of a year the net storage is usually very small and

$$r = E + \Delta f . \tag{7.3}$$

Summed over the whole globe, the horizontal redistribution of moisture always equals zero, since areas of runoff are exactly balanced by areas of inflow. Then, when the storage term may be neglected,

$$r = E\,. \tag{7.4}$$

The annual water balance also has this form over continental deserts, but not over oceanic deserts, where the annual excess of evaporation over precipitation is balanced by a net influx of water from other regions. Equation (7.4) does not hold, in general, for the monthly water balance of arid lands. Usually small amounts of moisture will be stored in the soil during the cool season and removed again in the warm season. This is true, for example, in southwestern Arizona, which has an average annual precipitation (and evaporation) of only 11.6 cm.

The latitudinal distribution of each component of the annual water balance of the earth's surface is shown in Figure 26. The data for precipitation and evaporation were taken from Table 1 of chapter 2. The runoff is calculated from equation (7.3). As indicated in chapter 2, precipitation exceeds evaporation poleward of 40° and between 10°N and 10°S. The resultant runoff from these regions, which is a maximum between 0 and 10°N (699 mm year^{-1}) and 50 and 60°S (526 mm year^{-1}), must replenish the excess evaporation of the subtropical latitudes (between 20 and 30°).

The excess precipitation of the northern hemisphere undoubtedly reaches the southern hemisphere via ocean currents, because only two major rivers, the Nile and the Amazon, cross the equator and both of these have their headwaters in the southern hemisphere.

The annual water balance of the various oceans and continents is given in Table 13. The data are taken from Budyko (1956), Budyko *et al.* (1962), and McDonald (1961). The Atlantic and Indian oceans are dry oceans, because they lose more water by evaporation than they gain by precipitation. The deficit is made up by inflow (negative runoff) from the surrounding lands (Δf_L) and from the Arctic and Pacific oceans (Δf_o). When converted into units of actual mass flow of water, the water exchange data given in the table show that the amount of water running off from the Arctic Ocean (0.16×10^{12} g sec^{-1}) equals the amount that flows into the Atlantic Ocean. Similarly, the quantity of water that runs off from the Pacific Ocean (0.68×10^{12} g sec^{-1}) is almost equal to the inflow into the Indian Ocean (0.70×10^{12} g sec^{-1}). Not too much significance should be attached to these results until better estimates of the components of the water balance are available for the oceans.

The oceans, as a whole, lose more water by evaporation than they gain by precipitation, the deficit being made up by runoff from the continents, over which precipitation exceeds evaporation. The world's wettest major geographical region is South America,

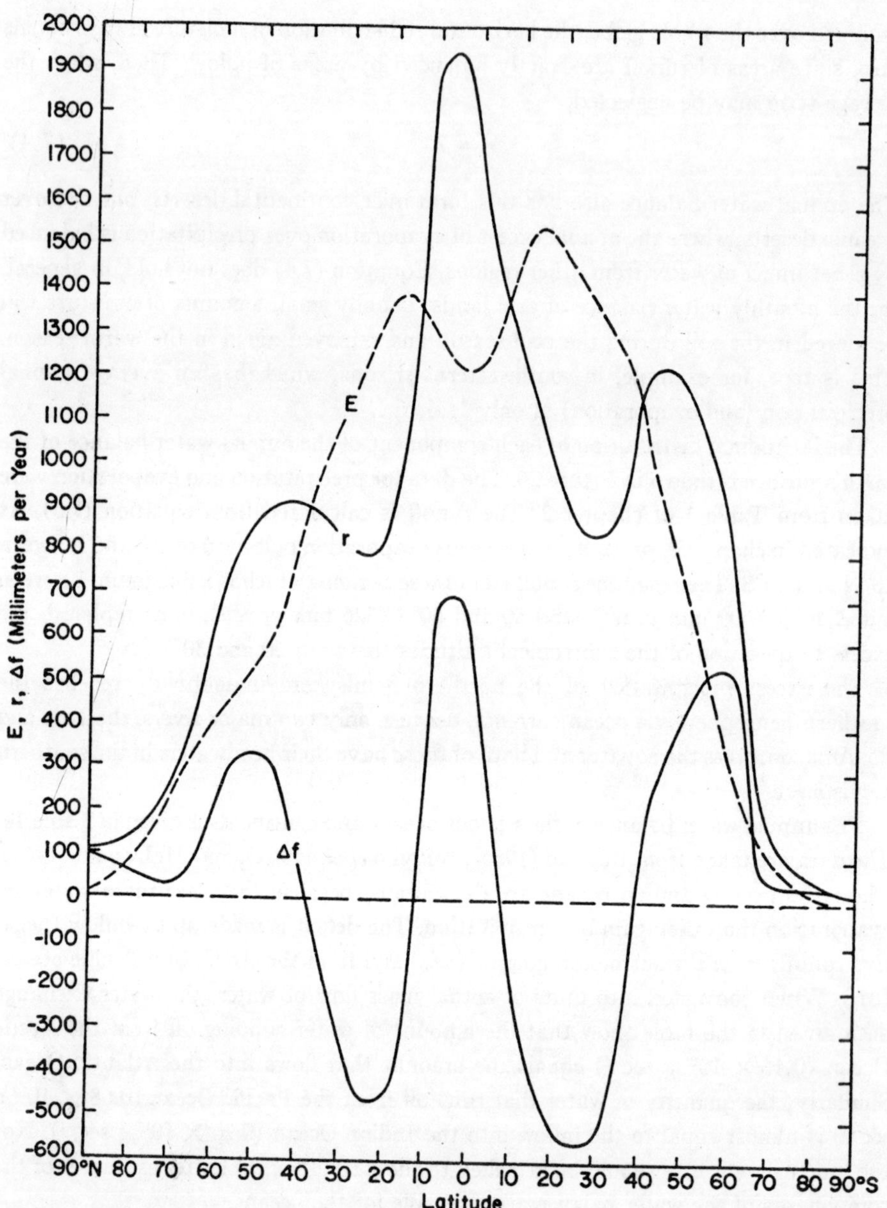

FIG. 26.—The average annual latitudinal distribution of evaporation (E), precipitation (r), and total runoff (Δf).

which annually receives 1,350 mm of precipitation. Almost two-thirds of this is lost to the atmosphere by evaporation. The remainder is discharged into the oceans at a rate of 0.27×10^{12} g sec^{-1}, which is exceeded only by runoff from the continent of Asia (0.31×10^{12} g sec^{-1}).

In general, the drier the continent the smaller the fraction of the mean annual precipitation that runs off. Thus, in Australia and Africa, both dry continents with a large proportion of their surface covered by deserts, more than three-fourths of the annual rainfall is lost by evaporation.

TABLE 13

ANNUAL WATER BALANCE OF THE OCEANS AND CONTINENTS

(mm year^{-1})

Ocean	E	r	Δf_L	Δf_o	Δf
Atlantic Ocean....	1,040	780	−200	− 60	−260
Indian Ocean......	1,380	1,010	− 70	−300	−370
Pacific Ocean......	1,140	1,210	− 60	130	70
Arctic Ocean......	120	240	−230	350	120
All oceans	1,250	1,120	−130	0	−130
Continent	E	r	Δf	$\Delta f/r$	
Europe...........	360	600	240	0.40	
Asia.............	390	610	220	0.36	
No. America......	400	670	270	0.40	
(United States)..	560	760	200	0.26	
So. America.......	860	1,350	490	0.36	
Africa...........	510	670	160	0.24	
Australia.........	410	470	60	0.13	
Antarctica........	0	30	30	1.00	
All land......	410	720	310	0.43	

The runoff ratio.—The apparent dependence of the runoff ratio $\Delta f/r$ on the mean annual precipitation may be explained in the following manner, rewriting equation (7.3) in the form

$$\frac{\Delta f}{r} = 1 - \frac{E}{r}. \tag{7.5}$$

For evaporation to occur, *both* a driving force and a source of energy for the phase transformation are required. The absence of either factor renders the other impotent. The driving force is the vapor pressure difference between the surface and the overlying air. Radiation is the main energy source. The first factor is very large in arid regions following the usually sporadic rainfall. The ground is moist and the air is relatively dry. Hence, in the presence of an adequate supply of energy, most precipitation evaporates before it has a chance to run off. In more humid regions the high air humidity limits the magnitude of the vapor pressure difference and leads to low evaporation rates and high runoff.

When the runoff ratio is plotted as a function of the mean annual precipitation for selected watersheds in the United States, a wide scattering of points results (Fig. 27). The scatter, however, is not random, since it is possible to group the data by geographical areas. Within each area the ratio a of the runoff ratio to the mean annual precipitation falls within rather narrow limits. The geographical distribution of a in the United States is shown in Figure 28 and summarized below. Both Δf and r are

Region	a (inches^{-1})
Rocky and Sierra Nevada mountains	≥0.02
Great Lakes, New England, northern Appalachians	0.01–0.02
East Central States and West Coast	0.005–0.01
Great Plains, Texas, Florida, and southwestern deserts	≤0.005

expressed in inches. Data for one hundred and fifty-two watersheds were used to construct the figure. The runoff and precipitation values were taken from reports by Kerr (1960) and the U.S. Weather Bureau (1959), respectively. For the most part, only watersheds 6,500 to 26,000 km^2 in area and with a runoff record of at least 40 years were considered. For comparison, data for twenty-one watersheds in central Europe, given by Schreiber (1904), are also plotted in Figure 27. These, in general, are compatible with the data for the northeastern United States.

For a given annual precipitation, the total runoff varies greatly across the country. A mean annual rainfall of 30 inches, for example, is accompanied by runoff of the order of 3 inches in Nebraska, 6 inches in Tennessee, 12 inches in New York, and 22 inches in the Rockies. These differences can be related to the seasonal distribution of precipitation and the radiation balance. Areas where the ratio a is large are, for the most part, areas of predominantly winter precipitation. Most of this runs off in the early spring when only limited radiative energy is available for evaporation. Areas where a is small, on the other hand, receive most of their precipitation in the late spring and summer when the evaporation potential is high.

One section of the country is omitted from the table above. This includes the Coast and Cascade mountains of Oregon and Washington, where practically all of the annual precipitation, ranging to 140 inches, runs off (Fig. 28).

Several attempts have been made to fit an all-inclusive curve to runoff-precipitation data. These include the relations

$$\frac{\Delta f}{r} = e^{-E_0/r} \qquad (7.6)$$

proposed by Schreiber (1904), and

$$\frac{\Delta f}{r} = 1 - \frac{E_0}{r} \tanh \frac{r}{E_0} \qquad (7.7)$$

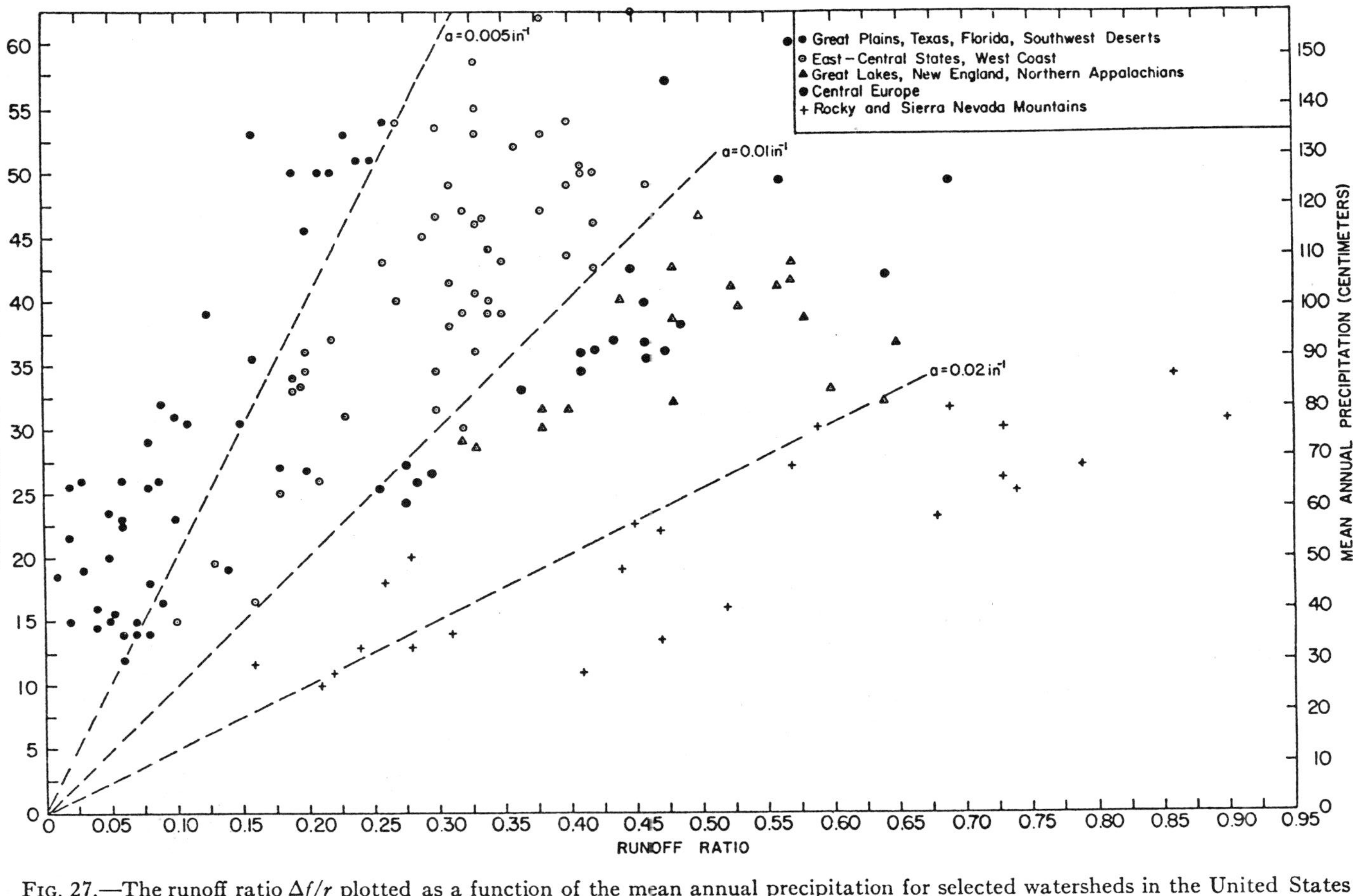

FIG. 27.—The runoff ratio $\Delta f/r$ plotted as a function of the mean annual precipitation for selected watersheds in the United States and Europe.

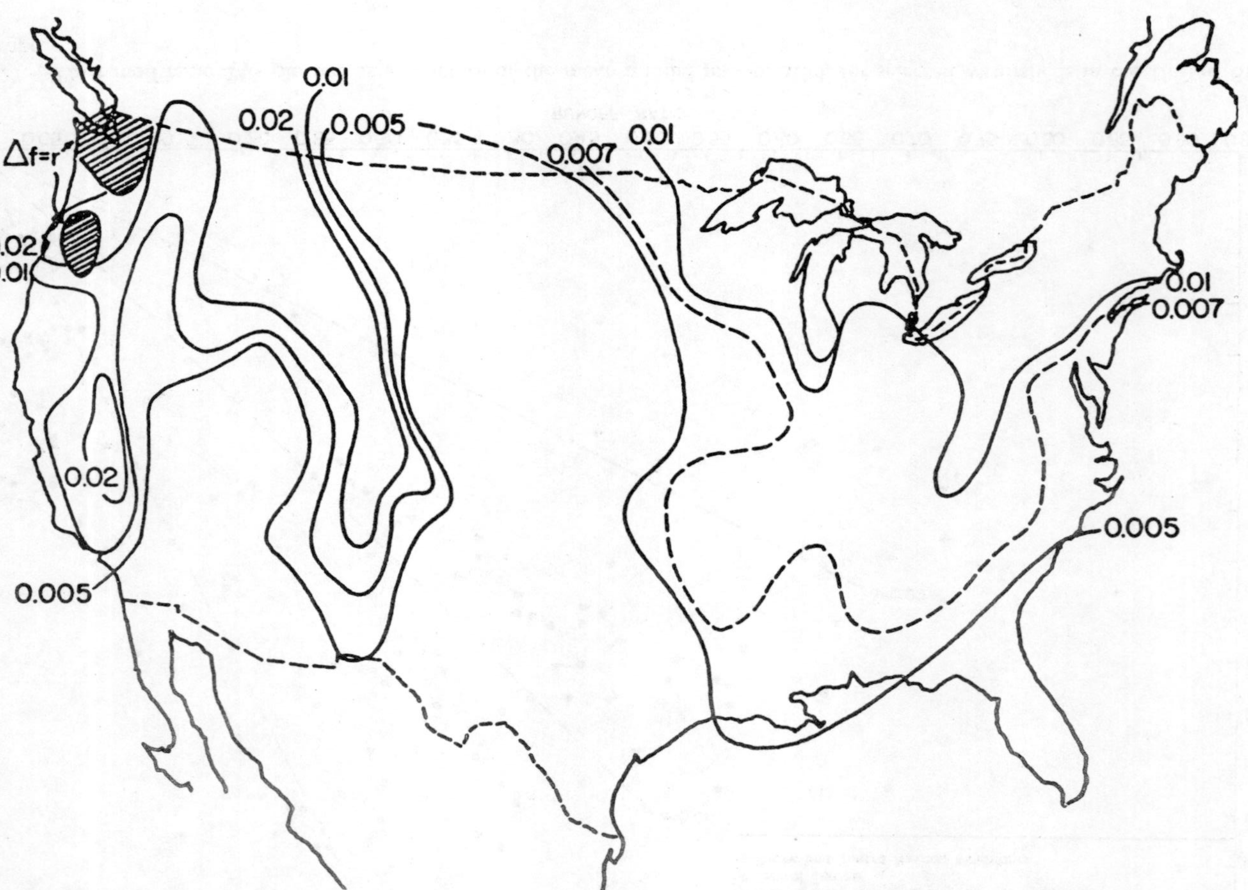

FIG. 28.—The distribution over the United States of the ratio of the runoff ratio $\Delta f/r$ to the mean annual precipitation r. Annual runoff and precipitation data for 152 watersheds ranging in size from 6,500 to 26,000 km² were used in this analysis. The units are inches^{-1}.

given by Ol'dekop (1911). Both of these satisfy the necessary conditions that $\Delta f \to 0$ as $r \to 0$ and $\Delta f \to r$ as $r \to \infty$. The quantity E_o is defined as the potential evaporation or evapotranspiration and represents the evaporation that would occur from a large, fully wetted surface; that is, one with an unlimited water supply. E_o is very closely related to the radiation balance R_o of the wet surface. In fact, if all available radiative energy is used for evaporation and there are no other energy sources,

$$E_o = R_o/L \tag{7.8}$$

where L is the latent heat of condensation. Budyko (1956) makes this assumption and takes the geometric mean of equations (7.6) and (7.7), solved for E/r from equation (7.5), and obtains

$$\frac{\Delta f}{r} = 1 - \left[x(1 - \cosh x + \sinh x) \tanh \frac{1}{x} \right]^{1/2} \tag{7.9}$$

where $x = R_o/Lr$. He claims (Monin and Budyko, 1964) that estimates of evaporation made with this equation, together with equation (7.5), for twelve hundred watersheds give values that average within 10 percent of those observed.

For the range of average annual precipitation normally encountered in the United States, the simple relation

$$\frac{\Delta f}{r} = ar \quad \text{or} \quad \Delta f = ar^2 \qquad \left(r \leq \frac{1}{a}\right) \tag{7.10}$$

where a is read from Figure 28, seems to be adequate. The coefficient a takes into account the annual variation of precipitation and spatial differences in the amount of energy available for evaporation. Equation (7.10) implies that a 10 percent increase in precipitation should be accompanied by a 21 percent increase in runoff, if a does not change. Actually, in arid regions, a usually increases with increasing precipitation, mainly because of an increase in the average soil-moisture content. Hence, a modest percentage increase in precipitation may result in a very significant increase in runoff.

The water balance as an index of climate.—Many climatic indexes that are based entirely or in part on the surface water balance are cited in the literature. Only a few will be mentioned here. Two of the most obvious are the runoff ratio $\Delta f/r$ and the evaporation ratio E/r, both of which have been used by Budyko (1956).

Budyko also uses the ratio R_o/Lr, which appears in equation (7.9) and which he calls the radiational index of dryness. Physically, it represents the ratio of the net amount of radiative energy available for evaporation from a wet surface (assumed to have an albedo of 0.18) to the amount of heat required to evaporate the mean annual precipitation. It is much greater than 1.0 in desert regions, where $R_o \gg Lr$, and less than 1.0 in humid regions, where $R_o < Lr$.

Perhaps it is more justifiable from a physical standpoint to use as a climatic index the ratio of the radiation balance R_o for a wet surface to the heat required to evaporate the mean annual precipitation less the mean annual runoff, that is, $R_o/L(r - \Delta f)$. This, of course, is simply equal to R_o/LE or to E_o/E, the ratio of potential to actual evapotranspiration.

Thornthwaite (1948) defines a quantity which he calls the moisture index I_m and which is very similar to Budyko's radiational index of dryness R_o/Lr. Thornthwaite writes

$$I_m = 100\left(\frac{\text{moisture surplus} - 0.6 \text{ moisture deficit}}{\text{moisture need}}\right)$$

where, averaged for the year, the surplus is equal to the runoff. The deficit is the difference between the amount of water which could be evaporated from a given surface if the supply was unlimited (potential evapotranspiration E_o) and the amount actually evaporated E. The need is the potential evapotranspiration E_o. In a recent series of publications, Thornthwaite and Mather (1962) drop the factor 0.6. The expression then becomes

$$I'_m = 100\left(\frac{\Delta f - E_o + E}{E_o}\right).$$

But, annually, $\Delta f = r - E$. Hence,

$$I'_m = 100\left(\frac{r}{E_o} - 1\right).$$

If all radiative energy available at a wet surface is used for evapotranspiration, then by equation (7.8), $LE_o = R_o$ and

$$I'_m = 100\left(\frac{Lr}{R_o} - 1\right).$$

Typical values of all the climatic indexes mentioned in this section are listed in Table 14 for various broad climatic types. The divisions are those suggested by

TABLE 14

SELECTED CLIMATIC INDEXES BASED ON THE WATER BALANCE

Climatic Region	$\Delta f/r$	E/r	R_o/Lr	E_o/E	I'_m	I_m
Tundra......	≥0.70	≤0.30	≤0.33	≤1.10	≥200	≥204
Forest.......	0.30–0.70	0.30–0.70	0.33–1.00	1.10–1.43	0–200	12–204
Steppe.......	0.10–0.30	0.70–0.90	1.00–2.00	1.43–2.22	−50–0	−28–12
Semi-desert....	0.03–0.10	0.90–0.97	2.00–3.00	2.22–3.09	−67–−50	−40–−28
Desert.......	≤0.03	≥0.97	≥3.00	≥3.09	≤−67	≤−40

Budyko (1956). This classification seems to fit observations in the United States very well and is in general agreement with the more detailed classification proposed by Thornthwaite and Mather (1955).

According to Budyko (1956), if two regions have nearly the same radiational index of dryness R_o/Lr, the growth of vegetation will be greatest in the one with the higher radiation balance for a moist surface R_o. This must occur because an increase in R_o results in a rapid increase in the potential evapotranspiration E_o. However, if R_o/Lr and, from Table 14, E_o/E are to remain fixed, the actual evaporation E including transpiration must also increase, thus increasing the production of vegetation. This conclusion could be modified slightly if the ratio of the discharge of water for transpiration to the weight accretion of dry mass in the vegetation is not constant for a given species. This ratio, or its inverse, is usually called the transpiration ratio. Some typical values from Lee (1942) are listed below.

Vegetation	Transpiration Ratio (g of water per g of dry leaf matter)
Corn	317
Wheat	375
Cotton	568
Alfalfa	626
Brome grass	880
Tumbleweed	272
Evergreen trees	140
Deciduous trees	825

Budyko (1956) also suggests that there is an optimum value of R_o/Lr near 1.0 at which, for a given R_o, productivity will be a maximum. He reasons that when the radiational index of dryness is close to this value there is just enough energy available to evaporate the mean annual rainfall. As a result, transpiration is not curtailed by a lack of moisture, as it would be if R_o/Lr were much greater than 1.0, nor is the soil waterlogged and poorly aerated, as it would be if R_o/Lr were much less than 1.0.

As a partial test of these two hypotheses, the radiation balance R_o and the radiational index of dryness R_o/Lr were determined with the aid of equation (7.9) for each of 152 watersheds in the United States. The values were averaged by states and compared with the annual timber growth as given by McArdle (1958) for each state for 1952. The results are summarized in the table on the following page. Each entry gives the average annual timber growth (m^3 km^{-2} of commercial forest) for the number of states given in parentheses and for the indicated ranges of R_o and R_o/Lr or $\Delta f/r$. For this analysis, the New England states were combined, as were New Jersey and Delaware.

When the relatively crude estimates made of runoff and precipitation in the forested land of each state are considered, the results are in good agreement with Budyko's hypotheses. The growth rate increases systematically with increasing radiation balance within each range of values of the radiational index of dryness and peaks at values of R_o/Lr between 0.8 and 1.0 in all but one of the R_o groups. The largest growth rates occur in extreme eastern Texas, Louisiana, Mississippi, Alabama, Georgia, and the Carolinas, where R_o exceeds 60 kly year^{-1} and R_o/Lr lies between 0.9 and 1.0. Missing from this group is Florida, which has a radiational index of 1.32 and a growth rate of only 149 m^3 km^{-2} year^{-1}. The lowest growth rates, less than 100 m^3 km^{-2} year^{-1}, are found in Arizona, New Mexico, Nevada, Utah, Wyoming, Nebraska, Colorado, and Montana. The first six of these states have values of R_o/Lr greater than 1.9; the last two have values of R_o less than 35 kly year^{-1}.

R_o/Lr	R_o (KLY YEAR^{-1})				$\Delta f/r$
	≤ 40	40–60	≥ 60	Mean	
≤ 0.8	212 (5)	207 (4)	259 (1)	215 (10)	≥ 0.4
0.8–1.0	204 (3)	216 (7)	263 (3)	224 (13)	0.3–0.4
1.0–1.9	84 (2)	88 (1)	206 (7)	169 (10)	0.1–0.3
≥ 1.9	....	56 (3)	138 (7)	113 (10)	≤ 0.1
Mean	184 (10)	173 (15)	192 (18)	183 (43)	....

THE WATER BALANCE OF THE ATMOSPHERE

The water balance of the earth's surface deals with only part of the hydrologic cycle. To complete the picture, the water balance of the atmosphere must also be considered. There are two sources of moisture for an atmospheric column extending from the surface to the top of the atmosphere. These are evaporation from the surface (E) and horizontal advection of vapor from the surrounding regions (c_i). Moisture is lost by precipitation (r), dewfall (D), and horizontal advection out of the column to surrounding regions (c_o). The rate of increase of moisture within the column is then

$$g_a = E + c_i - r - D - c_o$$

or

$$g_a = E - r - \Delta c\,, \tag{7.11}$$

again neglecting dewfall and letting $\Delta c = c_o - c_i$ be the net flux of moisture out of the column.

The flux divergence term Δc in equation (7.11) can be expressed in terms of the total precipitable water vapor w_a and the normal component u of the air velocity along the windward and leeward boundaries of an atmospheric column of horizontal surface area A. The appropriate relationship is

$$\Delta c = \frac{1}{A}\left[L_o(uw_a)_o - L_i(uw_a)_i\right]$$

or, since $\Delta c = c_o - c_i$,

$$c_o = \frac{L_o}{A}(uw_a)_o \quad \text{and} \quad c_i = \frac{L_i}{A}(uw_a)_i .$$

L_o and L_i are the lengths of the leeward and windward boundaries of the column. The product uw_a is usually obtained from a single atmospheric sounding by summing the product $(uw_a)_j$ over n tropospheric layers. In each layer both w_a and u are relatively constant. Measurements are rarely extended above 400 mb, because of the absence of appreciable water vapor at these heights.

Since the atmosphere can contain only relatively small quantities of water in any of its phases, the storage-rate term is usually much smaller than the other components of the balance in equation (7.11). The average annual value is close to zero; hence,

$$\Delta c = E - r . \tag{7.12}$$

On comparison with equation (7.3), which is the annual water balance equation for the earth's surface, it follows that

$$\Delta f = \Delta c .$$

Thus, the annual runoff from any region is just balanced by an influx of moisture into the air column above the region. As implied in Figure 26, water vapor is transferred from the subtropical atmosphere to higher and lower latitudes. To balance this loss there is an equal flow of water into the region at the surface.

The magnitude of the annual meridional flux of water vapor in the atmosphere across each latitude circle can be obtained from the values of Δf given in Figure 26 by assuming zero flux across the poles. The results, expressed in kilograms per year, are presented in Figure 29. Also shown are estimates made by Gabites (1950), Benton and Estoque (1954), and Starr, Peixoto, and Livadas (1958).

The data of Starr *et al.* (1958) are for the year 1950 and for the northern hemisphere only. They were obtained from all available meteorological soundings of wind and humidity by first determining the average annual value of uw_a at each of many stations and then averaging these along each latitude circle. Then representing the resulting values by $[\overline{uw_a}]$, where the bar indicates a time average and the brackets a latitudinal average, it follows that

$$[\bar{c}_m]A = L[\overline{uw_a}] . \tag{7.13}$$

Here, L is the length of the latitude circle and u is the meridional component of the air velocity. This method for estimating the meridional vapor flux $[\bar{c}_m]$ is subject to a number of shortcomings in the data and in the computation techniques, as detailed by Starr and White (1955). Nevertheless, it does provide an independent check of the more conventional estimates made with equation (7.12).

As is apparent from Figure 29, considerable uncertainty exists concerning both the magnitude and the sign of the vapor transfer, particularly between 25°N and 25°S.

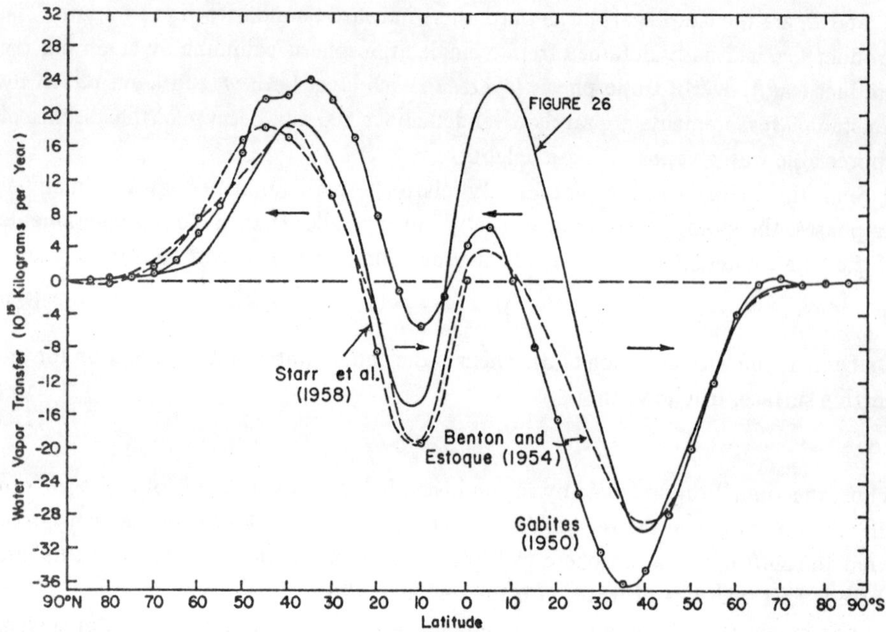

FIG. 29.—The latitudinal distribution of the average annual meridional transfer of water vapor in the atmosphere as derived from the Δf values of Figure 26. Also shown are estimates made by Gabites (1950), Benton and Estoque (1954), and Starr *et al.* (1958).

This uncertainty is due mainly to a lack of accurate knowledge of precipitation and evaporation rates in the tropics and subtropics. The presence of a strong northward vapor transfer centered between the equator and 15°S depends primarily on the magnitude of the difference between evaporation and precipitation in the two hemispheres. According to Budyko (1956), this difference is large, 120 mm year^{-1}, leading to a transfer rate across the equator almost twice as great as that shown by the solid line in the figure. The solid line is based on a difference $E - r$ of 64 mm year^{-1} (see Table 1) between the hemispheres.

In spite of the uncertainties, at least three features of the vapor transfer seem to be indisputable. First, the flux is almost surely poleward north of 20°N and south of 20°S. Second, the magnitude of the maximum between 35 and 40°S is apparently about 50 percent greater than the magnitude of that between 35 and 45°N. Third, there is a net flux of water vapor into the intertropical convergence zone from both the north and south.

The actual mechanism by which water vapor is transferred from one latitude to another has been studied by Starr and Peixoto (1964). Appendix 5 shows that the average vapor flux $[\bar{c}_m]$ across a given latitude circle can be expressed as the sum of three terms:

$$[\bar{c}_m] = [\bar{c}_m]_1 + [\bar{c}_m]_2 + [\bar{c}_m]_3 \,. \tag{7.14}$$

$[\bar{c}_m]_1$ represents the portion of the total flux associated with the existence of an average north-south tropospheric air motion across the given latitude. A non-zero annual value of $\bar{u}$ must be balanced by counterflow in the opposite direction in the upper troposphere and lower stratosphere, thus creating a cellular-type mean meridional circulation. $[\bar{c}_m]_2$ represents the flux associated with the large-scale semi-permanent cyclones and anticyclones of the general circulation. To the extent that these features can be looked upon as large turbulence elements embedded in the normal zonal motion of the atmosphere, this term represents a type of horizontal eddy transfer. Its existence depends on a non-zero correlation between the time-averaged values of w_a and u along the given latitude circle. $[\bar{c}_m]_3$ gives the flux due to the relatively small-scale traveling cyclones and anticyclones that appear on the daily weather map but are smoothed out of the annual average pattern. This is also an eddy flux and depends primarily on the daily values of u and w_a at a given station being correlated with each other. Thus, even though the average north-south motion at the station may be zero, this term can be large if daily winds from the south are characteristically associated with higher or lower precipitable water vapor than are winds from the north.

Values of each component in equation (7.14) multiplied by the appropriate surface area are listed in the table on the following page for selected latitudes in the northern hemisphere. These data were obtained by Starr and Peixoto (1964) and apply only to 1950. The units are 10^{15} kg year^{-1}. Negative values indicate a southward vapor transfer.

The data for $A[\bar{c}_m]_1$ suggest the existence of a three-cell meridional circulation, such as that described by Rossby (1945). At least for this single year the mean meridional flow in the lower and middle troposphere was northward between 35 and 55°N and southward at higher and lower latitudes. Only the southernmost cell, often called the Hadley cell, was important in the hydrologic cycle, transporting large quantities of water vapor equatorward between 5 and 25°N.

During 1950 both the large- and small-scale horizontal eddies of the general circulation carried water vapor northward at all latitudes in the northern hemisphere, with by far the greatest transport being associated with the migratory cyclones and anticyclones.

For many years vapor transfer in the atmosphere was considered of secondary importance in the hydrologic cycle. It was assumed, for example, that only the portion of the annual precipitation numerically equal to the annual runoff from a given region was actually derived from advected moisture. The rest was assumed to come from local evaporation. On the basis of these conclusions, it has been seriously proposed that precipitation in arid regions can be increased either by building many ponds and reservoirs or by planting a large number of water-consuming trees.

Latitude (°N)	$A[\bar{c}_m]$	$A[\bar{c}_m]_1$	$A[\bar{c}_m]_2$	$A[\bar{c}_m]_3$
80	− 0.4	....	....	....
70	1.7	− 0.7	0.2	2.2
60	7.4	− 0.4	1.0	6.8
50	16.8	2.6	0.6	13.6
45	18.3	3.0	0.5	14.8
40	17.2	3.4	0.9	12.9
30	9.2	− 3.2	3.2	9.2
20	− 8.5	−22.4	5.7	8.2
10	−19.4	−24.7	0.8	4.5
0	0.0	0.4	0.0	− 0.4

The correct role of advective moisture in the local water balance has been studied by Holzman (1937), Benton, Blackburn, and Snead (1950), and McDonald (1962) in the United States and by Drosdov (1956, 1961) and Budyko (1956) in Russia. Drosdov and Budyko present a simple theory that yields quantitative estimates of the importance of local and advected moisture in the precipitation regime of a particular region. A slightly modified version of their theory is derived below.

The total mass of water vapor advected into a given region of horizontal surface area A in unit time is equal to c_iA, or to uw_aL, where uw_a is the mean product of the precipitable water vapor w_a and the inflow speed u along the windward boundary of width L. In passing over the region, part of this vapor will condense and fall as precipitation and part will be stored in the atmosphere. If these parts are called r_aA and $g_{aa}A$, respectively, the rate at which moisture of external origin leaves the region is $(c_i - r_a - g_{aa})A$. This will be augmented by local evaporation, less storage and recondensation, to give for the total mass leaving the region in unit time,

$$c_oA = (c_i + E - r - g_a)A \; ,$$

in accord with equation (7.11).

Averaged over the whole region, the mass flow of vapor transported by the advected stream is

$$[c_i - \tfrac{1}{2}(r_a + g_{aa})]A$$

and by the total stream

$$[c_i - \tfrac{1}{2}(r + g_a - E)]A \ .$$

If it is now assumed that water vapor molecules of external and local origin have an equal probability of being stored or precipitated, the ratios r_a/r and g_{aa}/g_a should equal the corresponding ratio of the mass transfers. That is,

$$\frac{r_a}{r} = \frac{g_{aa}}{g_a} = \frac{c_i - \frac{1}{2}(r_a + g_{aa})}{c_i - \frac{1}{2}(r + g_a - E)}$$

or

$$\frac{r_a}{r} = \frac{g_{aa}}{g_a} = \left(1 + \frac{1}{2}\frac{E}{c_i}\right)^{-1} . \qquad (7.15)$$

Actually, as pointed out by Benton *et al.* (1950), the vapor molecules of local origin are initially concentrated close to the surface and may be more accessible for subsequent convective precipitation than the molecules of external origin. In this event, equation (7.15) overestimates r_a, but probably by not more than a few percent.

Two quantities are of interest. One is the proportion k_1 of the advected vapor flux that is precipitated, which is,

$$k_1 = \frac{r_a}{c_i} = \frac{2r_L}{E}$$

where r_L is the precipitation of local origin ($r_L = r - r_a$). The other quantity is the fraction k_2 of the total precipitation associated with the advected flux, which is

$$k_2 = \frac{r_a}{r} .$$

The theory has been applied to three regions with very different areas: the state of Arizona (excluding the Colorado River), European Russia, and the United States and Canada combined. The data for Russia were taken from Budyko (1956) and for the United States and Canada from Benton and Estoque (1954). The results are summarized in Table 15.

On the average only 11 percent of the moisture advected into Arizona is actually precipitated. The range is from about 3 percent in the spring, when the air is relatively dry and stable, to 20 percent in the summer (July to September), when afternoon convective showers occur almost daily in the tropical maritime air that covers most of the state. Advected moisture is by far the most important source of rainfall in Arizona, local sources contributing less than 20 mm to the annual total. Only in summer is the local contribution to precipitation greater than 10 mm. Apparently then, large reservoirs built along the windward borders of the state would have little effect on its climate, as suggested by McDonald (1962).

Even though it is difficult to look upon an essentially desert region as a major source of atmospheric water vapor, as proposed by Starr and Peixoto (1958), the state of Arizona is just that in the spring. The combination of very low precipitation rates

TABLE 15

ANALYSIS OF THE WATER BALANCE FOR SELECTED REGIONS
(Mm/3 months)

Region	Area (10^5 km²)	Measured Quantities	Winter	Spring	Summer	Fall	Year
Arizona..........	1.14	r	87	29	130	68	314
		E	38	116	126	33	313
		Δc	−49	72	1	−25	−1
		g_a	0	15	−5	−10	0
		k_1	0.14	0.03	0.20	0.12	0.11
		k_2	0.97	0.93	0.90	0.97	0.94
European Russia.	49.2	r	82	90	177	138	487
		E	15	96	143	40	294
		Δc	−66	−7	−35	−85	−193
		g_a	−1	13	1	−13	0
		k_1	0.69	0.30	0.34	0.43	0.39
		k_2	0.94	0.84	0.86	0.94	0.89
United States and Canada (1949)..	178	r	144	144	210	177	675
		E	58	125	253	125	561
		Δc	−90	−37	35	−35	−127
		g_a	4	18	8	−17	13
		k_1	0.55	0.60	0.67	0.70	0.63
		k_2	0.89	0.74	0.60	0.75	0.73

and high evaporation rates during this season leads to a marked increase in the moisture content of the air as it passes over the state. Part of this is stored, but most of it is advected to other regions. Since the annual runoff from Arizona is close to zero, the outflow of atmospheric moisture in the spring is balanced by inflow in the fall and winter.

European Russia is about fifty times larger than the state of Arizona. In an average year it receives almost 500 mm of precipitation, about 40 percent of which runs off. This loss is made up by an influx of moisture into the Russian atmosphere in all seasons, but especially in the fall and winter (December to February). Because the region is located in the latitude zone of frequent cyclonic storm activity and relatively high precipitation efficiency, a large portion of the advected moisture is precipitated, especially in winter when k_1 averages almost 0.70. In the other seasons the fraction drops appreciably but still averages nearly 0.40 for the year. Since evaporation rates are relatively low in European Russia, primarily because of its northern latitude, local moisture accounts for only slightly more than 10 percent of the total precipitation. In this respect, Russia and Arizona are similar.

The data for Canada and the United States are for the single year of 1949. The values given for Δc and g_a were obtained by Benton and Estoque (1954) from an analysis of atmospheric soundings at thirty radiosonde stations in the area. Evaporation rates were then determined as the residual in the water balance equation [eq. (7.11)]. These losses seem to be somewhat higher than might be expected, implying that only 17 percent of the total precipitation for the year either ran off or was stored in the soil. If these figures are correct, it follows from Table 15 that the continent was a moisture source in summer (June to August) and a sink in the other seasons. Between 55 and 70 percent of the water vapor advected over the region condensed and fell as precipitation. In spite of this high withdrawal rate and because of the large evaporation losses, an average of more than 25 percent of the total precipitation was of local origin, with the highest percentage, 40 percent, occurring in summer.

Benton *et al.* (1950) have also considered the annual water balance of the Mississippi River Basin, which covers almost half of the United States. According to their data, less than 20 percent of the moisture advected over the area is precipitated out. Yet this represents more than 90 percent of the total precipitation.

So it appears from these analyses that by far the most important source of local precipitation is advected moisture, even for areas of continental size. Local evaporation would seem to be of dominating importance in the precipitation regime of only those areas that are source regions for maritime tropical air and that are favored by relatively slight air motion. This would presumably include the oceanic subtropical highs, the intertropical convergence zone, and perhaps the Amazon Basin of South America.

8/ THE ENERGY BALANCE

All the heat (or energy) balance equations are based on the physical principle of the conservation of energy, just as the water balance equations are based on the conservation of matter. For this reason, it is sometimes appropriate to refer to them as energy conservation equations.

ENERGY BALANCE OF THE EARTH'S SURFACE

The energy balance equation for the earth's surface can be derived in exactly the same manner as the water balance equation. Consider a soil or water column extending from the surface to that depth where vertical heat exchange is negligible. (Annual temperature variations are noticeable to a depth of 5 to 22 m in soil and to 180 to 610 m in water; daily variations penetrate to about 1 m in soil and to at least 6 m in still water.) The net rate G at which the heat content of this column is changing is equal to the sum of the rates at which heat is being added by the absorption of solar radiation $(Q + q)(1 - a)$, by the absorption of longwave counter radiation from the atmosphere $I_{\downarrow}$, by the downward transfer of sensible heat from the air when the air is warmer than the surface $-H$, and by the horizontal transfer of heat into the column from the surroundings F_i, minus the rates at which heat is being lost by longwave radiation to the atmosphere $I_{\uparrow}$, by the transfer of sensible heat to the air when the air is cooler than the surface H, by evaporation LE where L is the latent heat of vaporization (about 590 cal g^{-1}), and by the horizontal transfer of heat out of the column F_o. Thus

$$G = (Q + q)(1 - a) + I_{\downarrow} - I_{\uparrow} - H - LE + F_i - F_o .$$

As indicated by equations (4.2) and (5.1), the first three terms on the right form the radiation balance R. It is customary, then, to write the equation

$$R = H + LE + G + \Delta F \tag{8.1}$$

where $\Delta F = F_o - F_i$ is the net subsurface flux of sensible heat out of the column. It is important only in the oceans and other water bodies where currents can transport considerable heat energy from one region to another. Over land ΔF is negligibly small and

$$R = H + LE + G\,, \tag{8.2}$$

which implies that the net available radiative energy is used to warm the air, evaporate water, and warm the soil.

At night there is no solar radiation and

$$R = -I = H + LE + G\,.$$

Since the effective outgoing radiation I is nearly always positive, the surface loses heat radiatively at night. To preserve a balance, sensible heat is usually transferred to the (cooler) surface from the (warmer) air (negative H) and from the (warmer) deeper soil layers (negative G). Negative evaporation (dew formation) also often occurs; however, in dry climates positive evaporation customarily continues all night, although at a much lower rate than during the day.

The storage term may be neglected in the annual heat balance equation, because energy stored in the oceans and continents during the spring and summer is all released in the fall and winter, except insofar as the oceans and continents are warming or cooling on a secular time scale. Thus, for oceans, $R = H + LE + \Delta F$, and for land, $R = H + LE$. The same equations apply to the daily energy balance, since the heat stored in soil and water in the morning and early afternoon is almost balanced by a heat loss in the late afternoon and at night.

Summed over the whole globe, ocean areas gaining heat by currents are always exactly balanced by those losing heat by currents. Thus, $\Delta F = 0$ and, when the storage term may be neglected,

$$R = H + LE\,.$$

The heat balance equation in the proper form applies over any time period, from one second to a million years. It is, however, an approximate equation to the extent that some small components have been neglected. These might be important locally at a particular time and include the following:

1. The melting of snow and ice in the spring, which requires about 2.4×10^{20} cal. If the melting occurs over a 100-day period on 70 percent of the land area between 40 and 60° in both hemispheres (22.6×10^6 km^2, or 4.45 percent of the earth's surface), this is equivalent to about 10 ly day^{-1}, which is between 5 and 10 percent of the radiation balance at these latitudes during the spring. When snow melt M is important, equation (8.2) should be written

$$R = H + LE + G + M\,.$$

2. The dissipation of mechanical energy of wind, waves, tides, and currents, which is estimated to lie between 1 and 10 ly day^{-1}. This is a source of heat energy and as such would appear on the left-hand side of the energy balance equation.

3. Heat transfer by precipitation, which is a source or sink of energy depending on whether the precipitation is warmer or colder than the underlying surface. This effect can normally be included in the storage rate term G. Its magnitude could easily reach 40 ly hour^{-1} during the brief but heavy showers that occur frequently on summer afternoons in many regions.

4. The expenditure of heat for photosynthesis, which is estimated to equal less than 8 percent of the net radiation and less than 5 percent of the total incoming solar radiation (Yocum, Allen, and Lemon, 1962; Lemon, 1962). Its average value is close to 1 percent of $Q + q$.

5. Gain of heat by the oxidation of biological substances. An average forest fire will give off about 850 ly day^{-1} of heat. This is roughly three times the net radiation received on a summer day over land.

6. Any surface item in Tables 2 and 3, including combustion, volcanic eruptions, earthquakes, street lighting, and the flux of heat from the earth's interior. According to Lee and MacDonald (1963), the latter quantity averages 0.132 ly day^{-1}, with maximum values of almost 0.7 ly day^{-1}.

The latitudinal distribution of the annual energy balance components for the oceans, continents, and the earth as a whole is given in Table 16. Most of these data are taken from Budyko *et al.* (1962). The values given for R and LE, averaged over land and water, are compatible with the values of R and E given in earlier chapters.

The net radiation is the only component of the energy balance to have the same latitudinal distribution over the oceans as over land. In both cases the maximum values are found in the tropics. Between 20°S and 30°N, the net radiation averages 115 kly year^{-1} over the oceans and 71 kly year^{-1} over land; the difference is explained by the relatively high surface albedo and temperature of the land areas, which are predominantly deserts. Thus, in this zone the absorbed solar radiation is less and the effective outgoing radiation is greater over land than over the oceans.

Poleward of 50° the net radiation is about the same over land and water. In these regions the albedo of water surfaces is relatively high, because of the low sun angle and the oceans are in many areas warmer than the land. Near the poles the net radiation is negative, the effective outgoing radiation exceeding the small amount of solar radiation absorbed by the ice- and snow-covered surfaces. In winter, mean values of zero net radiation are found as far south as northern New Mexico in the United States.

For the earth as a whole, the net radiation is about 70 percent greater over the oceans than over land.

On land the loss of heat for evaporation or the latent heat flux *LE* is greatest at the equator and decreases poleward. In the northern hemisphere a minimum occurs in the arid subtropical belt between 20 and 30°N. Farther north the heat loss for evaporation first increases to a secondary maximum between 30 and 50°N, the zone of

TABLE 16

MEAN LATITUDINAL VALUES OF THE COMPONENTS OF THE ENERGY BALANCE EQUATION FOR THE EARTH'S SURFACE

(kly year^{-1})

Latitude Zone	Oceans				Land			Earth			
	R	*LE*	*H*	Δ*F*	*R*	*LE*	*H*	*R*	*LE*	*H*	Δ*F*
80–90°N	...	...	...	...	...	...	...	−9	3	−10	−2
70–80	...	...	...	...	...	...	...	1	9	−1	−7
60–70	23	33	16	−26	20	14	6	21	20	10	−9
50–60	29	39	16	−26	30	19	11	30	28	14	−12
40–50	51	53	14	−16	45	24	21	48	38	17	−7
30–40	83	86	13	−16	60	23	37	73	59	24	−10
20–30	113	105	9	−1	69	20	49	96	73	24	−1
10–20	119	99	6	14	71	29	42	106	81	16	9
0–10	115	80	4	31	72	48	24	105	72	11	22
0–90°N	...	...	...	...	...	...	...	72	55	16	1
0–10°S	115	84	4	27	72	50	22	105	76	10	19
10–20	113	104	5	4	73	41	32	104	90	11	3
20–30	101	100	7	−6	70	28	42	94	83	16	−5
30–40	82	80	8	−6	62	28	34	80	74	11	−5
40–50	57	55	9	−7	41	21	20	56	53	10	−7
50–60	28	31	10	−13	31	20	11	28	31	11	−14
60–70	...	...	...	...	...	...	...	13	10	11	−8
70–80	...	...	...	...	...	...	...	−2	3	−4	−1
80–90	...	...	...	...	...	...	...	−11	0	−11	0
0–90°S	...	...	...	...	...	...	...	72	62	11	−1
Globe	82	74	8	0	49	25	24	72	59	13	0

frequent cyclonic storms, and then decreases again because of the inadequate heat supply.

In contrast to continental conditions, the loss of heat for evaporation over the oceans reaches a maximum, more than twice that reached over land, in the subtropics between 10 and 30°. Poleward of 30° a very potent energy source, present over water but not over land, is the large amount of heat transported from the tropics by ocean currents. As a result, in these latitudes the energy used for evaporation from the oceans far exceeds that used for evaporation from land and, in fact, even exceeds the radiation balance. For the earth as a whole, evaporation rates from land are only about one-third of those from the oceans.

The sensible heat flux or the turbulent heat exchange H increases steadily with latitude over the oceans and is considerably smaller than either the net radiation or the loss of heat for evaporation. The ratio of the sensible heat flux to the latent heat flux, H/LE, commonly referred to as the Bowen ratio, also increases with latitude, from about 0.05 near the equator to almost 0.50 at 70°N. Physically this change must be related to the poleward decrease of both air temperature and the associated saturation vapor pressure.

The sensible heat flux from the continents is greatest in the latitudes of the subtropical deserts (20 to 30°) and decreases markedly both poleward and equatorward. Its magnitude exceeds that of the heat loss for evaporation from 10 to 40°N and from 20 to 40°S. Poleward of 70° the earth's surface is normally colder than the overlying air; this implies a downward (or negative) sensible heat flux in these regions. On the icecaps this downward flux balances the radiational loss of energy. For the earth as a whole the transfer of sensible heat from the continents exceeds the transfer from the oceans by a factor of three.

Ocean currents in general transport sensible heat out of the zone between 20°N and 20°S, the maximum energy being taken up by currents slightly north of the equator. This heat is then transported to higher latitudes and is lost in greatest amounts between 50 and 70°N, where strong, warm ocean currents are especially active.

Table 16 shows that about 90 percent of the net radiation of all the oceans is used to evaporate water. The remaining 10 percent goes into warming the air by conduction and convection. On land these two forms of heat loss are almost equally important. For the earth as a whole, the loss of heat by evaporation accounts for 82 percent of the net radiation and turbulent heat exchange for 18 percent. Hence, the main method by which the radiative heat surplus of the earth's surface is dissipated and transferred vertically to the atmosphere is by evaporation of water. Eventually the resulting vapor will recondense, release heat, and partially offset the radiative energy deficit of the atmosphere.

The annual energy balance of the various oceans and continents is given in Table 17. The data are taken from Budyko (1963*a*). In Europe and North and South America most of the available radiative energy is used for evaporation. In Asia, Africa, and Australia, which contain about 90 percent of the world's deserts, most of the energy goes into warming the air.

The annual energy balance is about the same over the three major oceans. In each case about 90 percent of the available energy is used for evaporation. The small magnitude of ΔF suggests that the transport of heat by ocean currents does not have a great effect on the energy balance of each ocean as a whole. There is some indication of a transfer of heat from the Atlantic Ocean to the Arctic Ocean, but the magnitudes involved are probably within the margin of error of the calculations.

The annual variation of the energy balance at selected points in the United States and Europe and over the oceans is shown in Figures 30, 31, and 32. The data for Figures 30 and 32 were obtained by the methods outlined in Appendix 6. The data

TABLE 17

ANNUAL ENERGY BALANCE OF THE OCEANS AND CONTINENTS

(kly year^{-1})

Area	R	LE	H	ΔF	H/LE
Europe	39	24	15	0	0.62
Asia	47	22	25	0	1.14
North America	40	23	17	0	0.74
South America	70	45	25	0	0.56
Africa	68	26	42	0	1.61
Australia	70	22	48	0	2.18
Antarctica	−11	0	−11	0	....
All Land	49	25	24	0	0.96
Atlantic Ocean	82	72	8	2	0.11
Indian Ocean	85	77	7	1	0.09
Pacific Ocean	86	78	8	0	0.10
Arctic Ocean	− 4	5	− 5	−4	−1.00
All Oceans	82	74	8	0	0.11

for Figure 31 are based on direct measurements of the energy balance components by Aslyng (1960) over short grass near Copenhagen, Denmark, from 1956 to 1958, by Frankenberger (1960) over short grass near Hamburg, Germany, in 1957 and 1958, and by Koberg (1958) at Lake Mead, Arizona, in 1952 and 1953.

Except for minor irregularities, usually caused by increased cloudiness, such as in the northern Arabian Sea (Fig. 32) during the monsoon season (July and August), the net radiation increases uniformly from a minimum in winter to a maximum in summer over both land and water. The winter minimum is generally less than zero poleward of 40° and at high elevations in lower latitudes. The summer maximum normally lies between 300 and 400 ly day^{-1}. Values lower than this, oddly enough, are found mainly in the world's coolest and warmest regions, the latitudes poleward of 50° and the subtropical deserts. At high latitudes the net radiation is limited by

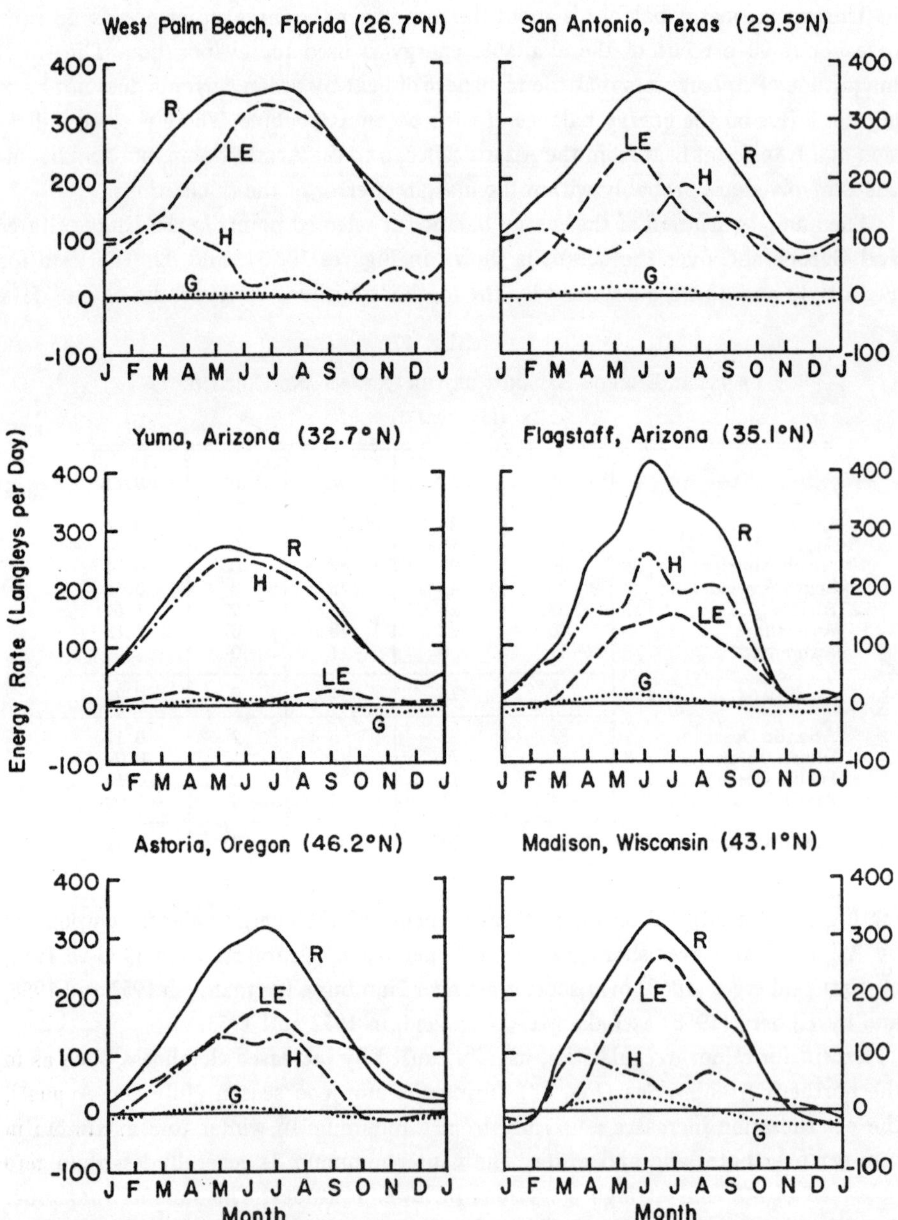

FIG. 30.—Average annual variation of the components of the surface energy balance at selected points in the United States.

the low sun angle and the high reflectivity of the surface, which is ice-covered in many areas. Deserts also have relatively high albedos, but, more important, because they are favored with clear skies and high surface temperatures, they radiate large quantities of energy to the atmosphere and to space. In midsummer the net radiation at Yuma, Arizona (Fig. 30), barely exceeds that at Copenhagen and Hamburg (Fig. 31).

Values of R greater than 400 ly day^{-1} in summer are found mainly over the relatively cloud-free portions of the subtropical oceans lying between 15 and 35° in both

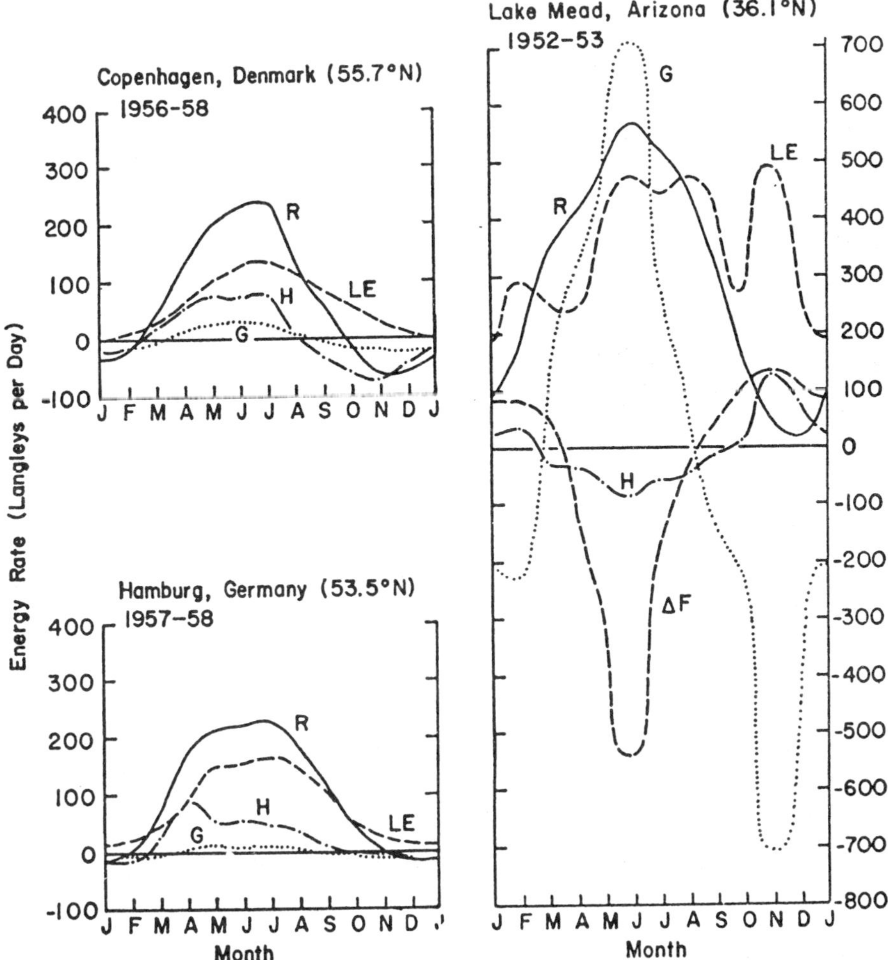

FIG. 31.—Average annual variation of the components of the surface energy balance near Copenhagen, Denmark, near Hamburg, Germany, and at Lake Mead, Arizona.

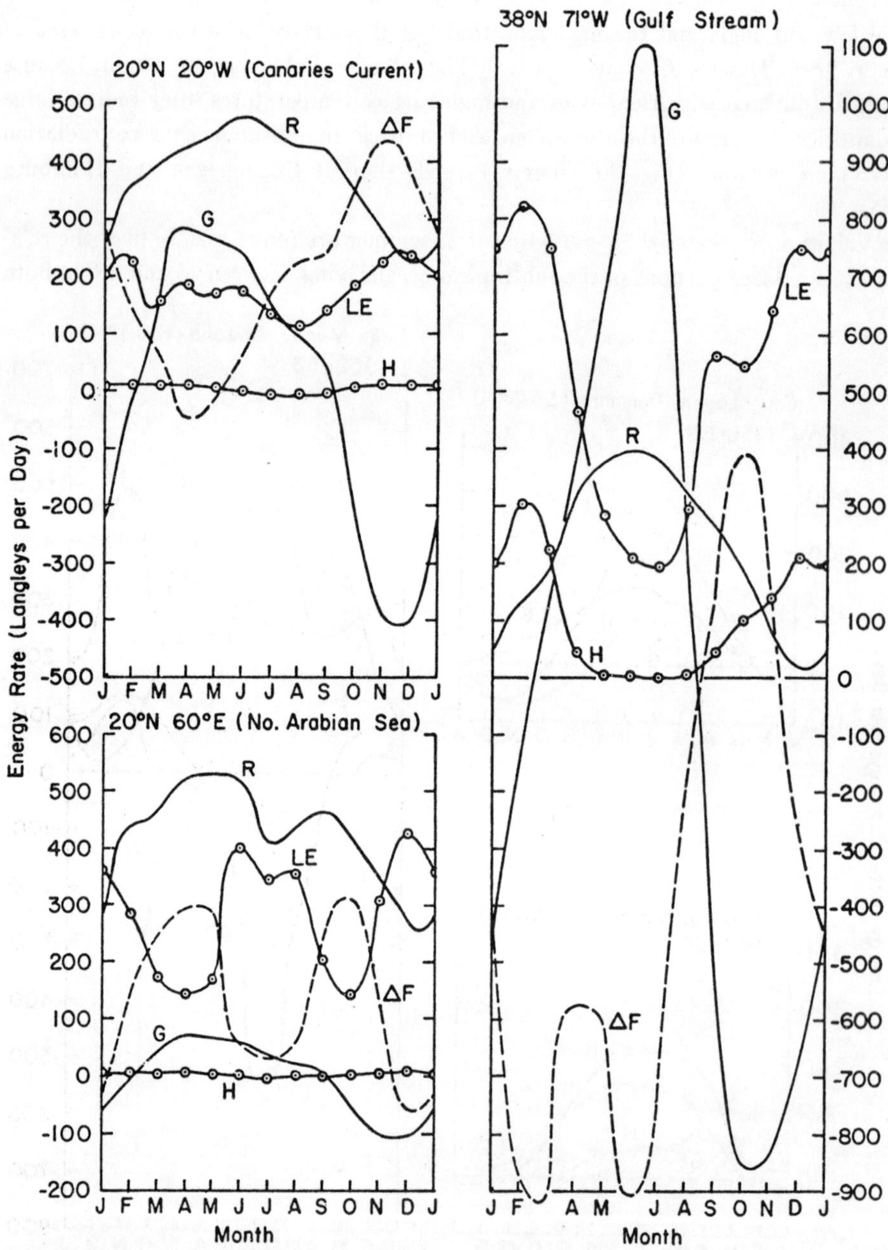

FIG. 32.—Average annual variation of the components of the surface energy balance at selected points over the oceans.

hemispheres. The annual radiation surplus is greater in the northern Arabian Sea (about 155 kly $year^{-1}$) than anywhere else in the world. This is a relatively humid region, with little cloudiness or precipitation (less than 10 cm $year^{-1}$), and moderate year-round sea surface temperatures (about 25°C).

Subtropical lakes and reservoirs, such as Lake Mead (Fig. 31) and the Salton Sea in the United States, have a positive radiation balance of more than 500 ly day^{-1} on clear summer days. These contrast quite sharply with the surrounding desert lands.

The latent heat flux LE or the evaporation rate E over land is mainly a function of the available radiative energy and the vapor pressure gradient between the soil and the overlying air. In the more humid regions the annual course of monthly evaporation is usually very similar to that of the net radiation, with a summer maximum and a winter minimum. All the stations in Figures 30 and 31 except Yuma display this pattern. At Yuma, in the driest section of the United States, evaporation is greatest in April, after the light winter rains, and in September, the wettest summer month.

Many continental areas with winter rains and summer drought, such as southern California and the lands bordering the Mediterranean Sea, have their greatest evaporation losses in the spring. Others, farther poleward, like Astoria, Oregon, continue to show a summer maximum, although considerably reduced in proportion to the energy available for evaporation. For example, the ratio LE/R is only 0.62 at Astoria, compared to 0.80 at West Palm Beach, Florida, which is considerably drier than Astoria but receives most of its precipitation during the warm season. At Astoria almost 70 percent of the annual precipitation runs off before it can evaporate.

At high latitudes in winter it is not unusual for the latent heat flux to be the only positive component of the energy balance. This occurs, for example, at both Copenhagen and Hamburg. In these cases the energy for evaporation and the energy to offset the negative radiation balance must come from the soil (negative G) and from the air (negative H).

Annual evaporation over land exceeds 1,000 mm in the Amazon Basin, parts of Central America, southern Florida, and some tropical islands of the South Pacific. Inland lakes and reservoirs in the subtropical deserts, such as Lake Mead, may lose more than 2,000 mm of water annually by evaporation.

Over the oceans the energy for evaporation is derived mainly from the water itself. As a result, the latent heat flux is poorly correlated with the net radiation and depends chiefly on the wind speed and the vapor pressure difference between the water surface and the free air. A wintertime evaporation maximum is not unusual, especially over the warm Gulf Stream and the Kurosiwo Current. Cold air flowing off the Asian and North American continents in winter is typically very dry and has a vapor pressure much lower than that of the warm water surface. Evaporation proceeds at an intense rate, drawing on the tremendous amount of energy stored in and advected by the ocean currents. In winter almost 400 mm of water is evaporated monthly from the Gulf Stream at 38°N 71°W (Fig. 32).

In summer the oceans are the primary air mass source and the flow is predominantly from water to land. As a result, both wind speeds and vapor pressure differences are reduced, which lead to decreased evaporation rates. An exception occurs in the northern Arabian Sea, which is influenced by the strong monsoon winds of summer and has a secondary evaporation maximum from June to August. The energy for this peak is apparently drawn from the net radiation (Fig. 32).

Most of the world's highest evaporation rates occur over the oceans. More than 2,000 mm of water is evaporated annually from parts of the Gulf Stream, the Kurosiwo Current, the central Indian Ocean, and the eastern South Pacific Ocean. Monthly totals in excess of 200 mm are not uncommon in winter in these areas. The northern Arabian Sea, with its extremely large radiative surplus, has a relatively low evaporation rate of the order of 1,500 mm year^{-1} because of the high air humidity. The excess energy is disposed of by ocean currents.

If all other factors are constant, the sensible heat flux from land surfaces to the atmosphere is greatest when and where the soil is driest and the temperature difference between the surface and the air is largest. Over deserts its annual course is similar to that of the radiation balance. Nearly all the available energy is used to heat the air. For this reason deserts are warm, in spite of their relatively low radiative surplus. At high latitudes in winter the surface, often snow-covered, is frequently colder than the overlying air and the sensible heat flux is directed downward.

Over the oceans the annual variation of the sensible heat flux to the atmosphere is quite similar to that of the latent heat flux but with much reduced magnitudes. It almost always has a positive maximum in winter when the water is warmer than the cold air flowing off the continents. In summer the sensible heat flux from the oceans is normally close to zero, or even slightly negative, especially over the cold currents and subtropical reservoirs.

The highest values of the sensible heat flux from the oceans in winter exceed 300 ly day^{-1} and are found over parts of the Gulf Stream and the Kurosiwo Current, in the Davis Strait between Canada and Greenland, in the Barents Sea, and in the Sea of Okhotsk. The role that these regions play in modifying air masses moving off the continents cannot be overestimated. For example, a column of air 1,500 m deep taking up the above amount of heat would be warmed by more than 8°C.

Although the magnitude of H over the warm currents in winter is comparable to the highest values observed over land in summer, the annual totals are much greater over land, exceeding 60 kly year^{-1} over the Sahara and Arabian deserts.

The flux of sensible heat into soil, G, is a function of the annual surface temperature range and the composition of the soil; the flux increases with the temperature range and the soil moisture content. Its magnitude, however, is always quite small. The maximum values, which occur in the spring, rarely exceed 30 ly day^{-1} or about 10 percent of the net radiation.

Over land, heat transfer into the soil is entirely by molecular conduction, whereas over water the currents themselves create a type of turbulent mixing which very efficiently transfers heat to depths of 100 m or more. As a result, the storage rate term G is quite important in those ocean areas and deep lakes and reservoirs where the surface waters undergo even a modest annual temperature variation. Water bodies in the northern hemisphere accumulate heat from March through August and release it from September through February. The variation is almost in phase with the variation of the net radiation.

The redistribution of heat by currents plays a very important role in the energy balance of oceans and lakes. In the summer and fall the cooler parts of the tropical and subtropical oceans, such as the Canaries Current, take up a large fraction of the radiative surplus, which is transported poleward, especially by warm currents. This radiative surplus ultimately is used for evaporation and for warming the air in winter at higher latitudes.

The interaction of the storage and transfer terms in the energy balance equation explains much of the observed annual variation of evaporation over water bodies. For example, at Lake Mead (Fig. 31), evaporation has a double maximum: in the early summer, when as much as 1,000 ly of energy is added to the lake daily by radiation and by the influx of warm water from the Colorado River, and in the late fall, when storage losses reach 700 ly day^{-1}. Over the Gulf Stream (Fig. 32) the annual course of evaporation is dominated by the storage rate term. In the spring and early summer most of the energy advected northward by the warm water to the south is stored and not used for evaporation until the fall and winter when the vapor pressure difference between the water surface and air is large and wind speeds are high. From September through November part of the stored energy is advected to higher latitudes.

The diurnal variation of the energy balance components over land in summer is very similar in many respects to the annual variation, as shown in Figure 33 for three selected points. Data are for an irrigated alfalfa-brome hayfield at Hancock, Wisconsin (Tanner and Pelton, 1960*b*), a small stand of irrigated Sudan grass at Tempe, Arizona (Van Bavel and Fritschen, 1964), and a barren dry lake near El Mirage, California (Vehrencamp, 1951, 1953). The alfalfa-brome and the Sudan grass were, respectively, 15.24 cm and 100 cm high. In each case the observations were made

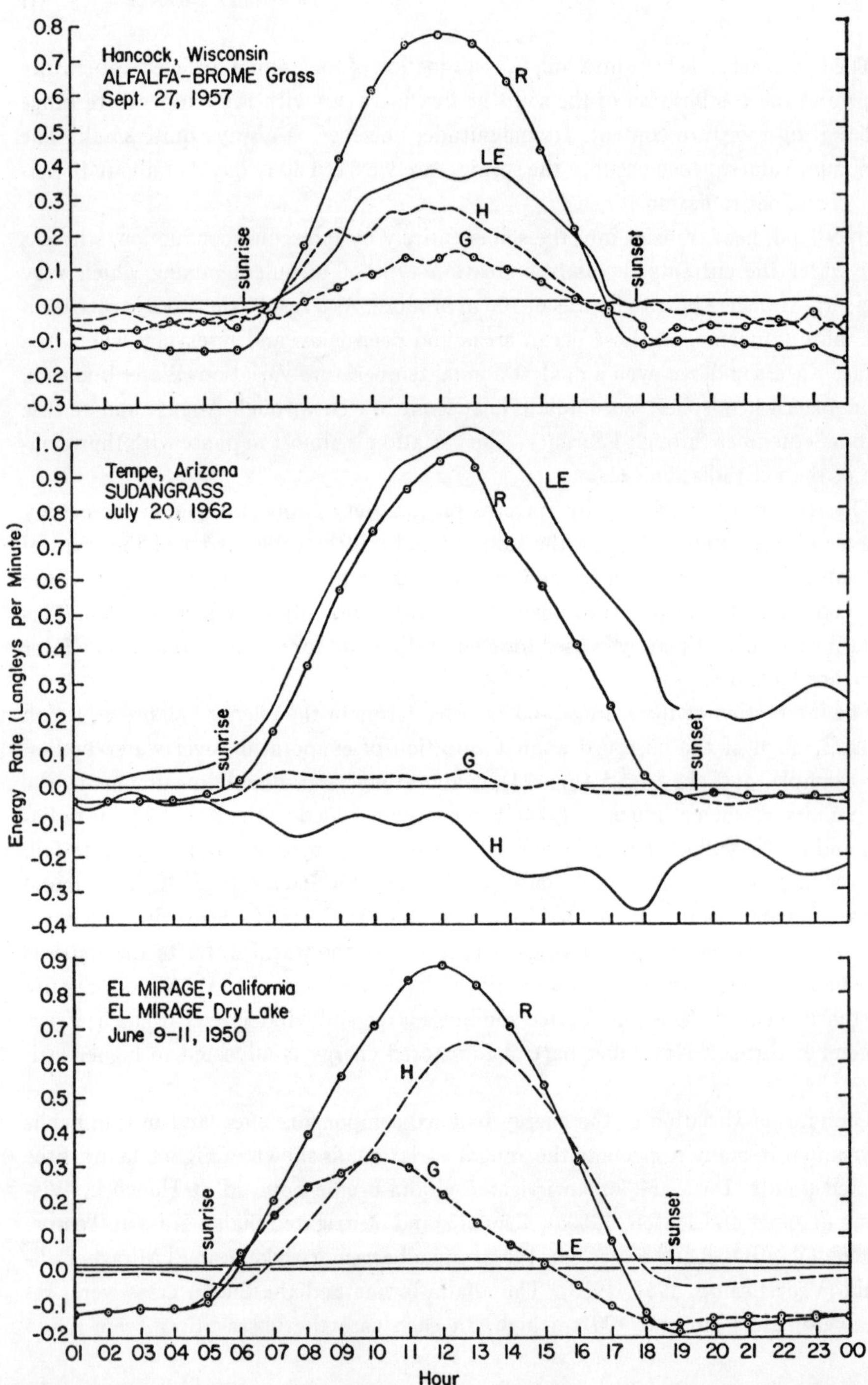

FIG. 33.—Average diurnal variation of the components of the surface energy balance over grass at Hancock, Wisconsin, and Tempe, Arizona, and over bare soil at El Mirage, California.

under clear skies. Air temperatures varied between 0 and 13°C at Hancock, between 25 and 45°C at Tempe, and between 17 and 29°C at El Mirage. Winds were light to moderate at each location, reaching a peak of less than 10 m sec^{-1} in the late afternoon at Tempe and El Mirage.

The net radiation varies systematically through the day taking on small negative values from about an hour before sunset until an hour after sunrise. At these times the slight incoming shortwave solar radiation just balances the effective outgoing longwave radiation. The daytime values increase rapidly to a maximum near noon which usually lies between 0.7 and 1.1 ly min^{-1} on clear days during the warmer half of the year. In temperate latitudes in winter the net radiation may be positive for only a few hours in the middle of the day.

Large differences exist in the magnitude of the latent heat flux LE at Hancock and Tempe, even though evaporation from the well-watered grass at both locations proceeds at close to the potential rate. The driving force for evaporation, the vapor pressure difference between the surface and the air, is much smaller at Hancock than at Tempe, where dry air from the arid surroundings is being continuously advected over the small stand of Sudan grass. As a result, the daytime evaporative energy demands at Hancock can be met quite adequately by the available radiative heat, but at Tempe this source must be supplemented with energy drawn from the air. This topic is discussed further in chapter 11.

At night the latent heat flux is very small and directed downward at Hancock; this indicates the formation of slight amounts of dew, especially in the early morning hours. Positive evaporation continues all night at Tempe. This seems to be typical of desert regions. The unusually large evaporation rates at Tempe in the early evening of July 20 were associated with an increase in the wind speed from about 3 m sec^{-1} at 1600 to almost 7 m sec^{-1} at 2000. At El Mirage the top soil layers contained only 2 percent moisture on a dry-weight basis during the period of observations and evaporation rates were negligible, totaling less than 0.5 mm for 24 hours.

The sensible heat flux H is normally directed upward during the day, when the temperature difference T_s-T between the surface and the overlying air is positive, and downward at night, when T_s-T is negative. For a given absolute value of T_s-T, the magnitude of H will be larger if the temperature decreases with height than if it increases with height, since warm air rises much more readily than it sinks. For this reason, the daytime positive values of the sensible heat flux at Hancock and El Mirage are much larger than the nighttime negative values. The maximum sensible heat loss to the air frequently occurs in the afternoon when wind speeds are normally stronger than they are earlier in the day.

The diurnal variation of the sensible heat flux over the Sudan grass at Tempe is atypical and occurs only in regions where a surface inversion persists all day. The rather steady negative increase of H from the early morning until the late afternoon parallels a similar increase in the wind speed. It is not uncommon for the sensible heat flux to be directed downward during the daytime over wet surfaces in arid regions (see the data for Lake Mead in Fig. 31).

The flux of sensible heat into the ground is often considered small. This is probably true over a 24-hour period, since energy gained by the soil during the day is almost balanced by energy lost at night. As a result, the ground temperature does not change appreciably from one day to the next. However, the heat used to warm the soil during the day is not negligible, varying on clear days from 5 to 15 percent of the net radiation for crops and grass to 25 to 30 percent of the net radiation for bare ground. On cloudy days the values may rise as high as 40 percent. The magnitude of this term rarely exceeds 0.3 ly min^{-1}, with the largest values over bare soil on clear days normally occurring one or two hours before noon, when the heating of the ground is greatest. The flow is reversed in the middle of the afternoon, about three or four hours before sunset, and remains negative until about an hour after sunrise.

ENERGY BALANCE OF THE EARTH-ATMOSPHERE SYSTEM

So far the discussion has dealt mainly with the heat balance of the earth's surface. It has been stated that radiative energy is used to warm the soil and air and to evaporate water. Over long periods of time, from years to decades, just enough energy is carried away from the surface in the form of latent and sensible heat to keep the surface from either warming up or cooling off. At the same time this energy transfer serves to overcome the net radiative cooling of the atmosphere, thus establishing an equilibrium state for the whole system.

In chapter 5 it was noted that there must be a poleward energy transfer in order to eliminate the net radiative energy surplus of the tropics and subtropics and the energy deficit of middle and high latitudes. The magnitude of the required transfer in kcal year^{-1} was shown in Figure 19 and is given again in the bottom part of Figure 34. The actual mechanisms by which this transfer is accomplished can be determined by considering the energy balance of the earth-atmosphere system.

A column extending from the top of the atmosphere to that depth in the soil or water where the vertical heat exchange is negligible may be heated by four mechanisms: (1) by the absorption of solar radiation $Q_s(1 - a_g)$, where a_g is the planetary albedo; (2) by the condensation of water vapor, approximately Lr; (3) by the horizontal flux of sensible heat into the column by air currents C_i; and (4) by the hori-

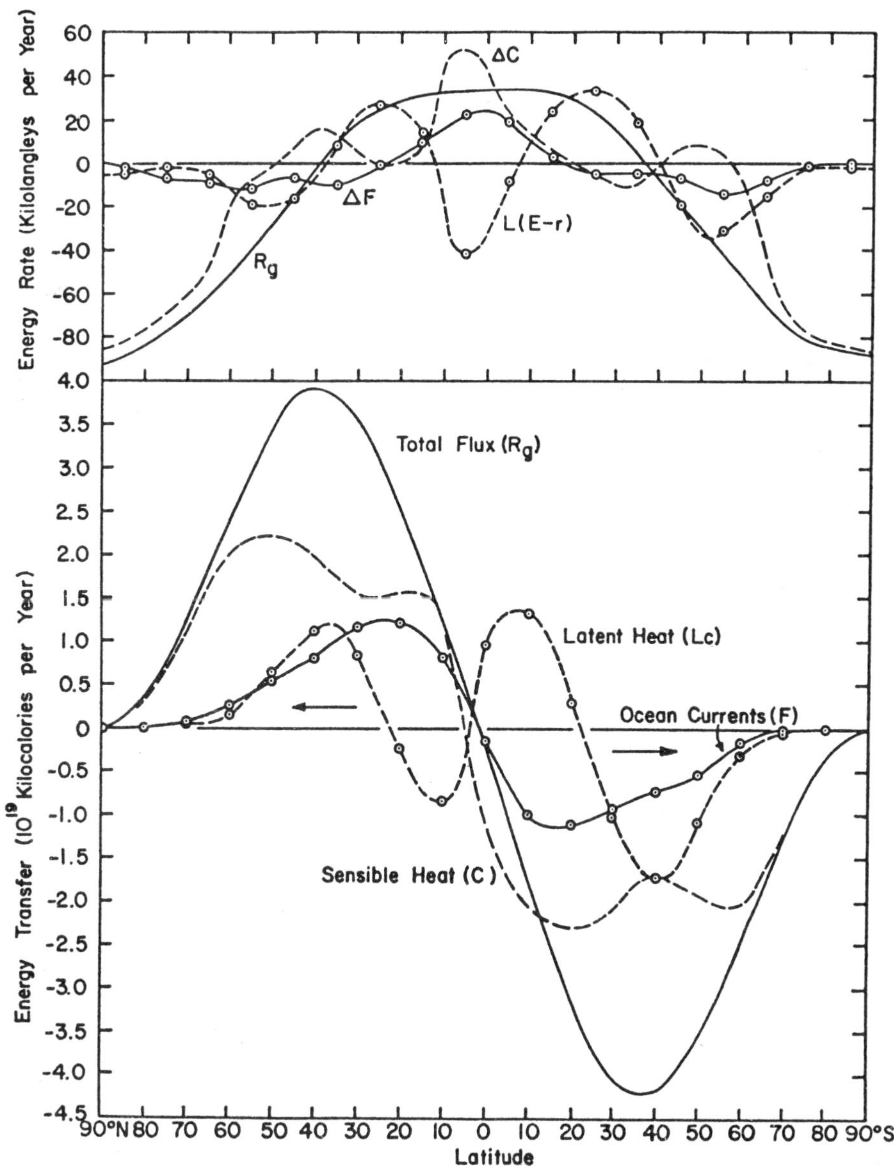

FIG. 34.—The average annual latitudinal distribution of the components of the energy balance of the earth-atmosphere system in kly year^{-1} (*top*) and of the components of the poleward energy flux in 10^{19} kcal year^{-1} (*bottom*).

zontal flux of sensible heat into the column by ocean currents F_i. The column may be cooled (1) by longwave radiation to space I_g; (2) by evaporation of water, approximately LE; (3) by the horizontal flux of sensible heat out of the column by air currents C_o; and (4) by the horizontal flux of sensible heat out of the column by ocean currents F_o. If G_g represents the net heating rate, it follows that

$$G_g = Q_s(1 - a_g) + Lr + C_i + F_i - I_g - LE - C_o - F_o .$$

The difference $Q_s(1 - a_g) - I_g$ is, by definition (chap. 5), the radiation balance of the column R_g. If

$$\Delta C = C_o - C_i \quad \text{and} \quad \Delta F = F_o - F_i$$

are the net flux of heat out of the column by atmospheric currents and ocean currents, respectively, the above equation reduces to

$$R_g = L(E - r) + G_g + \Delta C + \Delta F . \tag{8.3}$$

Over the period of a year the net warming or cooling of the column will be very small. Hence, G_g may be neglected in comparison to the remaining terms in the annual energy balance equation for the earth-atmosphere system. Then

$$R_g = L(E - r) + \Delta C + \Delta F , \tag{8.4}$$

which over land, where ΔF is zero, becomes

$$R_g = L(E - r) + \Delta C .$$

Annually for the whole globe each term is zero, since regions of energy inflow are exactly balanced by regions of energy outflow.

The latitudinal distribution of the components of the annual energy balance of the earth-atmosphere system is shown in the top part of Figure 34. The radiation balance is taken from Figure 19, ΔF from Table 16, and $L(E - r)$ from the values of Δf in Figure 26. The atmospheric flux ΔC is the residual term in equation (8.4).

The processes by which the radiation surplus between 40°N and 40°S is transported poleward can be deduced from the latitudinal distribution of the other energy balance components. The regions north of 50°N and south of 60°S and the subtropical zones between 20 and 30°N and 20 and 40°S gain energy by the meridional transfer of atmospheric sensible heat from lower latitudes. The major source region lies between the equator and 10°N. This zone has been called the atmosphere's firebox by Malkus (1962), because it is here that almost half of the atmosphere's sensible heat originates, most of it coming from the release of latent heat in the stormy intertropical convergence zone.

To obtain the magnitude of the meridional sensible heat transfer in the atmosphere, the ΔC values can be integrated with the boundary condition that there is no flux across the poles ($C_o = 0$ for the latitude belt from 80 to 90°). The results are shown in the bottom part of Figure 34. The transfer is northward north of 5°N and southward south of 5°N and accounts for roughly 60 percent of the total meridional heat transfer. This is significant since more than 80 percent of the vertical energy flux from the earth's surface to the atmosphere is in the form of latent heat of water vapor and less than 20 percent occurs as a vertical transfer of sensible heat. Apparently, then, most of the evaporated water must recondense within the same latitude zone in which it originates, the released energy being carried poleward as sensible heat.

The meridional transfer of sensible heat has a double maximum in both hemispheres, one in the subtropics between 15 and 25° and the other in high middle latitudes between 50 and 60°. The first is located on the poleward side of the tropical rain belt and the second on the poleward side of the most intense cyclonic activity.

As noted earlier, ocean currents in general transport sensible heat out of the zone between 20°N and 20°S, the energy eventually being used for evaporation and warming the air in higher latitudes. The magnitude of the transfer is greatest between 10 and 30° in both hemispheres, as indicated in the bottom part of Figure 34.

Earlier estimates suggested that ocean currents could account for, at most, about 10 percent of the total meridional heat transfer; however, the more recent data presented here indicate a value closer to 25 percent in the northern hemisphere and 20 percent in the southern hemisphere. Even though most of this transfer occurs in rather low latitudes, it could still be an important climatic factor for the whole globe.

Before attaching too much significance to the above result, we should note that there is still a great deal of uncertainty concerning the role of ocean currents in the meridional heat transfer. For example, Albrecht (1961), in a thorough study of the energy and water balances of the oceans, arrives at a latitudinal distribution of ΔF that would give a northward transfer at all latitudes except 10, 60, and 70°S, with a maximum of 1.12×10^{19} kcal year^{-1} at 20°N.

As pointed out in chapter 7, the difference between the annual evaporation E and the annual precipitation r represents the net moisture flux out of an atmospheric column. Hence,

$$L(E - r) = L(\Delta c)$$

represents the net flux of latent heat out of the column. The meridional distribution of the latent heat transfer, shown in the bottom part of Figure 34, is similar to the meridional distribution of the water vapor transfer, shown in Figure 29, and subject to the same uncertainties.

The latent heat flux accounts for only 20 percent of the total meridional energy transfer in the northern hemisphere and 25 percent of the total transfer in the southern hemisphere. In the tropical zone between 10°N and 10°S the meridional latent and sensible heat transfers are in opposite directions and of similar magnitudes. Water vapor is carried into the zone at low levels by the cool, moist northeast trades of the northern hemisphere and the southeast trades of the southern hemisphere, thus providing the necessary moisture for the heavy precipitation of the intertropical convergence zone. A significant portion of the latent heat released is then carried to higher latitudes as sensible heat. The mechanism for the sensible heat transfer is not well understood. The transfer could be accomplished either by a mean poleward-directed air flow at relatively low levels above the trade-wind inversion or by tropical disturbances.

The results of Figure 34 are summarized in the table below, which gives the percentage of the total poleward energy flux attributed to sensible heat (C), latent heat (Lc), and ocean currents (F) at 30, 50, and 70° in the northern and southern hemispheres.

Latitude	*C*		*Lc*		*F*	
	NH	*SH*	*NH*	*SH*	*NH*	*SH*
30°........	43	52	24	25	33	23
50°........	65	54	18	31	17	15
70°........	89	96	4	3	7	1

ENERGY BALANCE OF THE ATMOSPHERE

The total temperature change ΔT experienced by an air column extending from the ground to the top of the atmosphere is equal to the sum of the changes due to the absorption of shortwave solar radiation ΔT_Q, the emission of longwave radiation ΔT_I, the vertical transfer of sensible heat from the ground ΔT_H, the net horizontal transfer of sensible heat into the column from the surroundings ΔT_C, and the net release of latent heat ΔT_L. The effect of evaporation of raindrops before they reach the ground must be included in ΔT_L. Thus,

$$\Delta T = \Delta T_Q + \Delta T_I + \Delta T_H + \Delta T_C + \Delta T_L \,. \qquad (8.5)$$

This relationship may be transformed directly into an energy balance equation by using equation (3.11) or (3.12). When this is done, the terms represent, from left to right, the net heat storage rate G_a; the heating rate due to the absorption of solar radiation $Q_s(1 - a_a)$, where a_a is the albedo of the atmosphere; the negative of the effective outgoing radiation from the atmosphere I_a; the sensible heat flux from the ground H; the net horizontal flux of sensible heat into the column from the surroundings $-\Delta C$; and the product of the latent heat of condensation L and the net rate at which water vapor is condensed within the column, approximately the precipitation rate r. Hence, in complete analogy with equation (8.5),

$$G_a = Q_s(1 - a_a) - I_a + H - \Delta C + Lr \tag{8.6}$$

or, since

$$R_a = Q_s(1 - a_a) - I_a\,,$$

$$R_a = \Delta C - Lr - H + G_a\,. \tag{8.7}$$

The same result could be obtained by recalling that

$$R_g = R + R_a$$

and subtracting the heat balance equation for the earth's surface, equation (8.1), from that for the earth-atmosphere system, equation (8.3), letting

$$G_g = G + G_a\,.$$

The storage rate term may be neglected in the annual energy balance equation for the atmosphere. Then

$$0 = \Delta T_Q + \Delta T_I + \Delta T_H + \Delta T_C + \Delta T_L$$

or

$$R_a = \Delta C - Lr - H\,.$$

The average annual latitudinal distribution of each component is shown in Figure 35. Units of both kly year^{-1} and °C day^{-1} are used. $Q_s(1 - a_a)$ is taken from Figure 7, in which it equals $C_a + A_a$; I_a is taken from Figure 14; H is taken from Table 16; ΔC is taken from Figure 34; and Lr is taken from Figure 26.

The balance component that we are most uncertain about is the flux divergence term ΔC, which was obtained as the residual in the energy balance equation and, hence, contains the cumulative errors of the other components. In the following table the values of ΔC given in Figure 35 in units of kilolangleys per year are compared with those obtained by Vinnikov (from Budyko and Kondratiev, 1964), Davis (1963), Peixoto (1960), Mintz (1954), and Gabites (1950) for selected latitude belts in the northern hemisphere. The data obtained by Peixoto and Mintz are for the single years of 1950 and 1949, respectively, and refer only to the migratory eddy component of the total flux divergence. Davis attributes the differences between his results and the results of these two investigators to the existence of a significant transport by mean meridional motions of the type described in chapter 7.

In general, the agreement among the various sets of data is good and suggests that the values of ΔC obtained are accurate to at least 20 kly year^{-1} or about 0.2°C day^{-1}. In the southern hemisphere and at higher latitudes in the northern hemisphere, the data of Vinnikov and Gabites agree in all latitude zones to within 10 kly year^{-1} with the values read from Figure 35.

Warming of the atmosphere by the absorption of shortwave solar radiation averages about 0.5°C day^{-1}, ranging from 0.7°C day^{-1} in the humid tropics to 0.2°C day^{-1} near the poles (Roach, 1961). At all latitudes this warming is less than half of the longwave cooling, which reaches a maximum of 1.5°C day^{-1} at the equator and averages 1.3°C day^{-1}. The net result is that the atmosphere is cooling radiatively at all

Latitude Zone	Fig. 35	Vinnikov (1964)	Davis (1963)	Peixoto (1960)	Mintz (1954)	Gabites (1950)
60–70°N....	−47	−32	−32	−31	−53	−36*
50–60.......	− 8	− 2	−12	−14	−15	− 8
40–50.......	7	3	− 6	9	9	0
30–40.......	12	− 1	−11	17	33	11
20–30.......	−1	−14	−19	8	18	5
10–20.......	8	4	....	....	....	10
0–10.......	52	52	....	....	....	39

* These values include sensible heat transfer by ocean currents.

latitudes at a rate of about 0.8°C day^{-1}. To balance this cooling, warming must occur either by convection, advection, or the release of latent heat.

Poleward of 60°N and 60°S most warming occurs by the advection of warmer air from lower latitudes; this emphasizes the importance of the general circulation in maintaining the energy balance of the earth. The ultimate source of this heat seems to be the narrow tropical zone between the equator and 10°N, which cools 0.6°C day^{-1} as a result of the advection to higher latitudes of about half of the latent heat released by condensation within the zone.

Data presented by Rubin and Weyant (1963) and Rubin (1964) indicate that in the latitudes of the Antarctic continent most sensible heat transport across the antarctic boundary is accomplished through a mean meridional circulation, with poleward flow above 600 mb and equatorward flow below. Less than one-fourth of the heat transport is attributed to horizontal eddies, principally to migratory cyclones and anticyclones.

Direct heating of the atmosphere by a transfer of sensible heat from the surface is quite small, exceeding 0.2°C day^{-1} only in the dry subtropics of the northern hemisphere. Most atmospheric warming between 60°N and 60°S is due to the release of latent heat by condensation. This heating is strong enough between the equator and

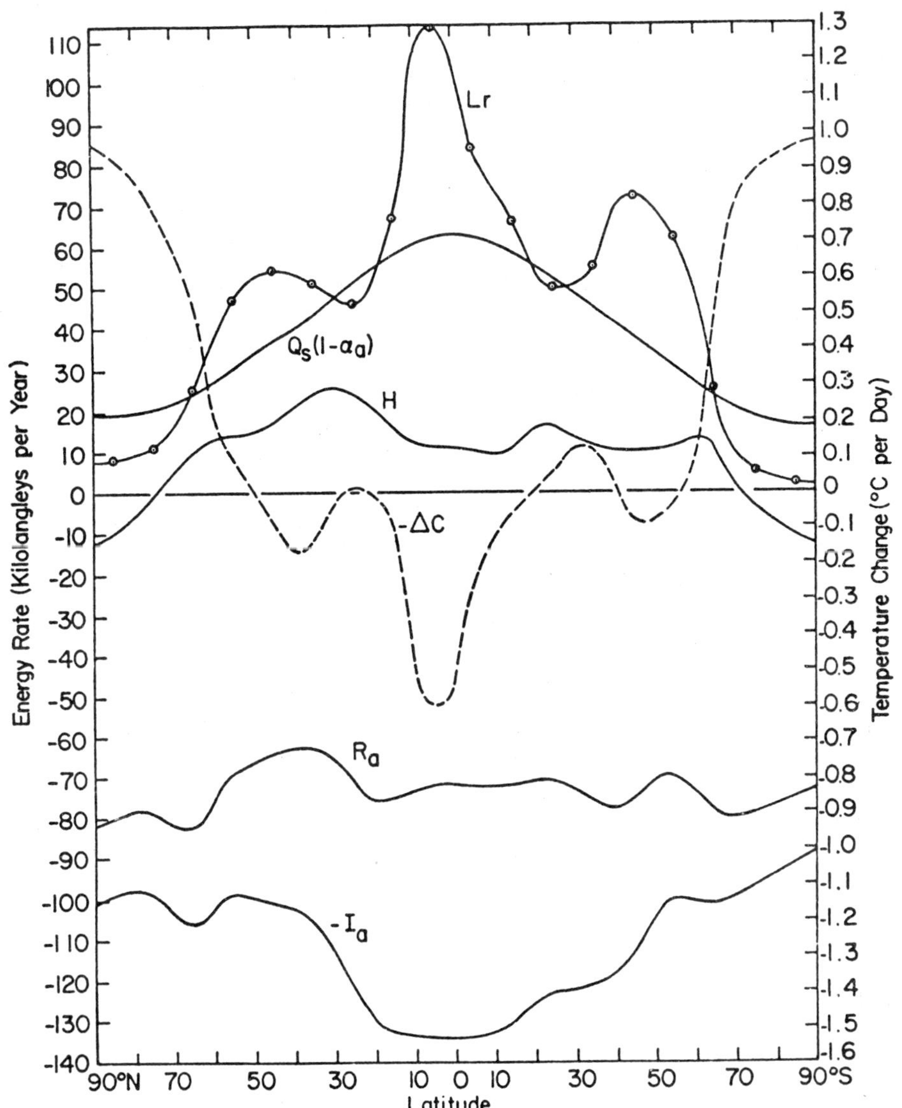

FIG. 35.—The average annual latitudinal distribution of the components of the energy balance of the atmosphere.

10°N to offset the cooling by radiation and by the advection of sensible heat to other latitudes.

The fact that the atmosphere poleward of 60° is warmed primarily by a meridional flux of sensible heat that has its ultimate source in the tropics suggests that these two regions are climatically very closely interrelated. It may be reasonably supposed, that a decrease in the sensible heat transfer poleward across 60° is accompanied, first, by lower temperatures at these high latitudes and, second, by a corresponding decrease in the sensible heat transfer out of the tropics.

The latter could be accomplished either by a decrease in tropical precipitation or by a weakening of the meridional circulation. Most likely both of these factors work together, since the moisture for precipitation is fed into the intertropical convergence zone primarily by the meridional trade-wind circulation. The change in the mean precipitation of the latitude zone between the equator and 10°N required to produce significant warming or cooling poleward of 60° is likely to be very small, of the order of 1 cm per 1.6°C temperature change according to Sellers (1964*a*), and not readily detectable from presently available data. Besides reducing polar temperatures and tropical precipitation, a decrease in the north-south air motion would perhaps also lead to higher equatorial temperatures and a strengthened zonal circulation.

ENERGY AND WATER BALANCES OF AN AIR LAYER CLOSE TO THE GROUND

Since the following three chapters deal mainly with physical processes taking place at or near the earth's surface, it is appropriate at this point to discuss briefly the energy and water balances of an air column extending a short distance Δz above the ground. Equation (8.7) for the energy balance and equation (7.11) for the water balance again apply. Here, however, additional terms must be added in each case to take into account the vertical transfer of energy through the top of the column. Thus, in accord with Figure 36, the energy balance is

$$G_a = R_1 + H + C_i + Lr - R - H_1 - C_o$$

or

$$G_a = \Delta R + \Delta H - \Delta C + Lr , \qquad (8.8)$$

and the water balance is

$$g_a = E + c_i - E_1 - c_o - r$$

or

$$g_a = \Delta E - \Delta c - r . \qquad (8.9)$$

In these equations H_1, R_1, and E_1 are the vertical fluxes of sensible heat, radiative energy, and water vapor, respectively, through the top of the column. Also

$$\Delta R = R_1 - R\,, \quad \Delta H = H - H_1\,, \quad \text{and} \quad \Delta E = E - E_1\,.$$

The difference ΔR is equivalent to R_a, since it represents the net amount of radiative energy absorbed by the column. If the column is extended to the top of the atmosphere, $R_1 = R_g$ and $H_1 = E_1 = 0$, the equations then reducing to their usual forms.

The advection components, ΔC and Δc, of equations (8.8) and (8.9) are related to the mean wind speed $\bar{u}$, the horizontal gradients of the mean temperature $\bar{T}$ and the

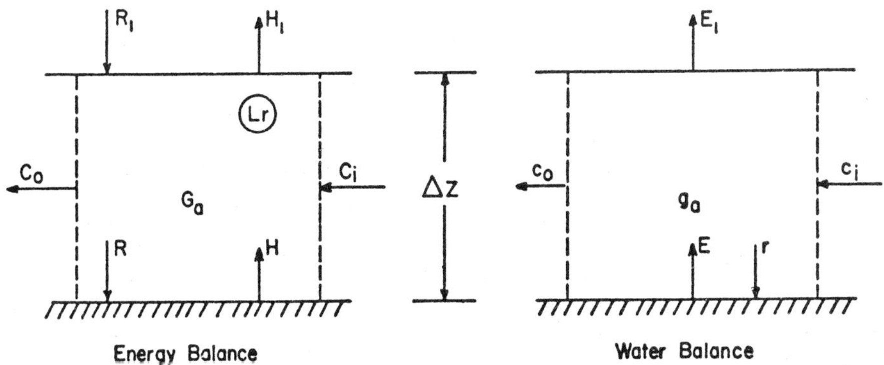

FIG. 36.—The energy and water balances of an air column extending a short distance Δz above the ground. Each component is considered positive when directed as shown.

mean specific humidity $\bar{q}$ in the direction x of the mean wind, and the depth Δz of the air column. Thus,

$$\Delta C = \bar{\rho}\, c_p \bar{u}\, \frac{\Delta \bar{T}}{\Delta x}\, \Delta z \quad \text{and} \quad \Delta c = \bar{\rho} \bar{u}\, \frac{\Delta \bar{q}}{\Delta x}\, \Delta z\,, \tag{8.10}$$

if the wind velocity does not vary horizontally. $\bar{\rho}$ and c_p are, respectively, the mean air density and the specific heat of air at constant pressure.

In accord with equation (3.11), the two storage rate terms, G_a and g_a are given by,

$$G_a = \bar{\rho}\, c_p\, \frac{\Delta \bar{T}}{\Delta t}\, \Delta z \quad \text{and} \quad g_a = \bar{\rho}\, \frac{\Delta \bar{q}}{\Delta t}\, \Delta z\,. \tag{8.11}$$

Typical magnitudes of the components of equations (8.8) and (8.9), and also of the surface energy balance, are given for a number of locations and times in Table 18. The values are based on equations (8.10) and (8.11) and on data given by Crawford and Dyer (1962), Rider and Robinson (1951), Funk (1960), Kraus (1958), Priestley (1959), and Lettau and Davidson (1957). All components, including those of the water balance (multiplied by $L = 585$ cal g^{-1}), are given in millilangleys per minute and refer to an air layer 1 m thick.

The first set of data in Table 18 refers to the windward edge of an irrigated field in a dry climate. Large horizontal gradients of temperature and moisture exist here, with the drier, warmer air upwind. As a result, both advection terms are large and of the same magnitude as the net radiation. The vertical fluxes of both sensible and latent heat change rapidly with height. In this particular case, H is directed downward at the surface and upward at 1 m. The downward flux at the surface provides extra energy for evaporation, in excess of that available from the net radiation.

TABLE 18

TYPICAL MAGNITUDES OF THE COMPONENTS OF EQUATIONS (8.8) AND (8.9) AT VARIOUS LOCATIONS AND TIMES

(Energy units of mly per minute)

Location	Energy Balance (air) (mly min^{-1})					Water Balance (air) (mly min^{-1})				Energy Balance (sfc) (mly min^{-1})			
	G_a	ΔR	ΔH	ΔC	Lr	g_a	ΔE	Δc	r	R	H	LE	G
Windward edge of an irrigated field (midmorning in summer)	2	10	−426	−418	0	1	372	371	0	643	−258	743	158
Center of large dry field (midday in summer)	1	18	− 17	0	0	−1	− 1	0	0	546	204	194	148
Center of large field (early evening of clear night)	−1	− 7	6	0	0	0	0	0	0	− 60	− 63	35	− 32
Forest clearing (early evening at time of fog formation)	−2	− 1	− 2	0	1	−1	− 1	0	0	− 62	− 22	− 8	− 32

Under ideal conditions, at the center of a field with an area of at least 10^5 m^2, the horizontal gradients of temperature and moisture are zero or very small and the advection terms vanish. Whether such conditions exist often is questionable. For example, with $\bar{\rho} = 10^{-3}$ g cm^{-3}, $c_p = 0.24$ cal g^{-1} °C^{-1}, $\bar{u} = 100$ cm sec^{-1}, $L = 585$ cal g^{-1}, and $\Delta z = 100$ cm, it would take a temperature gradient of only 1.4°C per 100 m and a specific humidity gradient of only 0.57 g kg^{-1} per 100 m to give values of ΔC and $L\Delta c$ equal to 20 mly min^{-1}.

When the advection terms can be neglected, the vertical fluxes of sensible and latent heat change slowly with height, the ratios $\Delta H/H$ and $\Delta E/E$ for the air layer from the surface to 1 m usually equaling less than 0.10 to 0.15, as shown for the second and third sets of data in Table 18. In these cases, the approximations $H = H_1$

and $LE = LE_1$ are within the accuracy to which the energy balance components can be measured by existing methods. Thus, in the absence of appreciable advection effects, the vertical fluxes of sensible and latent heat from the earth's surface can be estimated from measurements made a short distance above the ground, as outlined in chapter 10. Since, from equations (8.10) and (8.11), both the advection and storage rate terms increase with increasing Δz, the above approximations usually become poorer as the depth of the air column increases.

During the daytime the net radiation at 1 m is usually larger than that at the surface; this implies a positive value for ΔR in equation (8.8). However, the difference is small, of the order of 1 to 5 percent of the net radiation, and the approximation $R = R_1$ is quite acceptable for practically all conditions. In spite of its small magnitude, the convergence of radiative energy into the volume during the day can lead to appreciable warming. For example, a value of ΔR equal to 18 mly min^{-1} in a 1-m air layer is equivalent to a heating rate of 45°C hour^{-1}. Most of this is due to the emission and absorption of infrared radiation, since the absorption of solar energy by the air layer is extremely small and could not possibly account for a radiative temperature change of much more than 1°C hour^{-1}. Hence, it is permissible to write for both day and night

$$\Delta R = R_1 - R = I - I_1 = \Delta I$$

and for the energy balance,

$$G_a = \Delta I + \Delta H - \Delta C + Lr\,.$$

The above heating rate is much larger than that ever observed and must be almost balanced by cooling by vertical convection and possibly horizontal advection. Thus, as mentioned in chapter 4, in the absence of advection effects, almost as fast as an air layer near the ground is warmed radiatively, energy is carried to higher levels by convection.

At night or when the temperature increases with height, surface air is cooled radiatively, again in accord with the results of chapter 4. Since the cooling rate is often considerably larger than that observed, warming primarily by vertical convection must also occur. That is, warm air is brought from higher levels to the surface layers by turbulent mixing. At night the magnitude of ΔR can, on occasion, exceed 10 percent of the net radiation, making the approximation $R = R_1$ less exact than during the day.

The above pattern of radiative warming and convective cooling during the day and radiative cooling and convective warming at night extends, with decreasing magnitude, to a height of about 10 m. According to Elliott (1964), when the effects of horizontal advection can be neglected, the ratios of the radiative and convective temperature changes, ΔT_I and ΔT_H, respectively, to the observed temperature change ΔT, vary with height both day and night in the following manner:

Height	$\Delta T_I/\Delta T$	$\Delta T_H/\Delta T$
Below 30 cm.....	>30	<−29
1 m..............	8–10	−9–−7
10–12 m..........	1	0
100 m............	0.1	0.9

At all levels the radiative temperature change has the same sign as the observed change, positive during the day and negative at night. The ratio $\Delta T_H/\Delta T$, on the other hand, is negative below 10 to 12 m, where convective and radiative processes tend to cancel one another, and positive above 10 to 12 m, where the two processes work together to produce the observed temperature change.

Elliott also concluded from his analysis of the data of Lettau and Davidson (1957), taken over mown prairie grass in Nebraska, that the vertical flux of sensible heat is within 20 percent of its surface value at all levels below 200 to 500 m. This, of course, will be true only over very homogeneous terrain where horizontal temperature gradients are negligible.

The last set of data in Table 18, taken from Kraus (1958), refers to conditions in a forest clearing on October 12, 1956, at 1730, near the time of formation of ground fog. This is the only type of situation where the term Lr in equation (8.8) is likely to be important. The latent heat released by condensation warmed the air 0.2 to 2.5°C, if the water vapor density was 10^{-7} to 10^{-6} g cm^{-3}. This warming was more than offset by radiative and convective cooling, which together produced an observed temperature drop of 8.5°C between 1700 and 1800. At the same time that the fog was forming, moisture was being withdrawn from the cooling air layer and deposited as dew on the grass surface, as indicated by the negative values of both the latent heat flux LE and the vertical divergence of the vapor transfer ΔE.

9/ HEAT TRANSFER IN SOIL

The rate at which heat flows through a soil level at a depth z below the surface is directly proportional to the vertical temperature gradient existing at that level. Thus, at the depth z_1,

$$G_1 = -\lambda_1 \left(\frac{\Delta T}{\Delta z}\right)_1 \tag{9.1}$$

where the heat flow is positive and downward when the temperature decreases with depth. The constant of proportionality λ is called the thermal conductivity. According to Chang (1958), this constant is a function of the composition, moisture content, and temperature of the soil. Physically, the thermal conductivity represents the rate at which heat energy passes through unit area of a given substance when a temperature gradient of 1°C cm^{-1} exists. A good insulator has a low thermal conductivity.

De Vries (1963) lists the following values for the thermal conductivity of various soil constituents. In general, λ is highest for a soil containing abundant quartz and

Substance	T(°C)	λ(mcal cm^{-1} sec^{-1} °C^{-1}) 1 mcal = 10^{-3} cal
Quartz	10	21
Clay minerals	10	7
Organic matter	10	0.6
Water	10	1.37
Ice	0	5.2
Air	10	0.06

least for a soil rich in organic matter. It increases with increasing moisture content, since water replaces air in the soil pore spaces, and is higher for frozen soil than for unfrozen soil. Some of these relationships are shown in Figure 37, in which the dependence of the thermal conductivity on the volume fraction of water in the soil (x_w) is given for Fairbanks sand at a temperature of 4.4°C, for quartz sand at 20°C, and for clay and peat at 18°C. The data for the two sands are taken from material presented by De Vries (1963) and based on a very complete study by Kersten (1949). The data for the clay and peat soils are from information presented by Van Wijk (1963). The volume fraction of mineral and organic matter in each soil type is given by x_s.

The thermal conductivity of sand and clay soils increases very rapidly with increasing moisture content at low values of x_w. These soils become relatively poor insulators if even a little water is added.

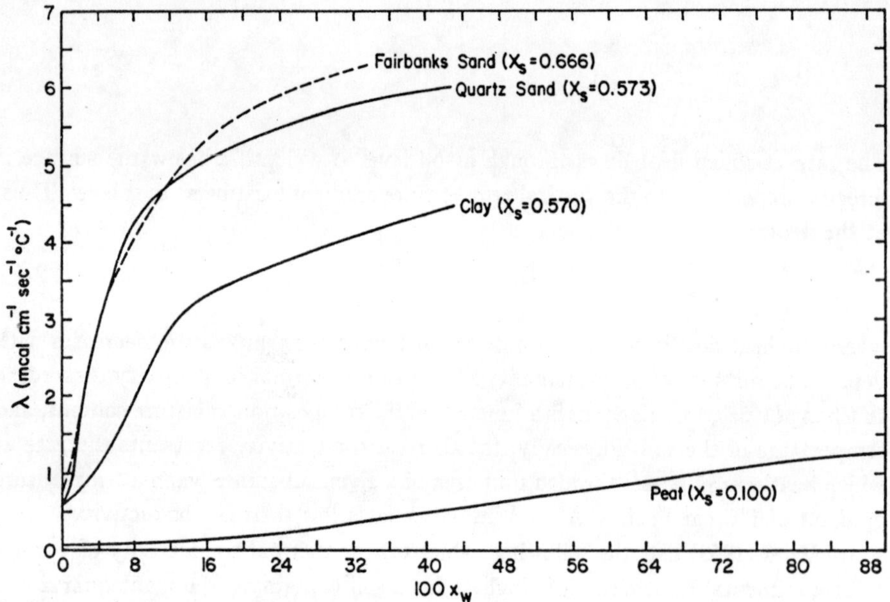

FIG. 37.—Dependence of the thermal conductivity λ on the volume fraction of water x_w for four different soil types.

In the absence of horizontal temperature gradients, the rate at which a soil layer is warming or cooling can be determined with the aid of equation (9.1). Thus, if

$$G_1 = -\lambda_1 \left(\frac{\Delta T}{\Delta z}\right)_1$$

is the vertical flux through the top of the layer and

$$G_2 = -\lambda_2 \left(\frac{\Delta T}{\Delta z}\right)_2$$

is the flux through the bottom of the layer, then

$$\Delta G = G_2 - G_1 = \lambda_1 \left(\frac{\Delta T}{\Delta z}\right)_1 - \lambda_2 \left(\frac{\Delta T}{\Delta z}\right)_2 \qquad (9.2)$$

and represents the net flux of heat energy out of the layer. This, by equation (3.11), is directly proportional to the observed time rate of temperature change,

$$\Delta G = -C \frac{\Delta T}{\Delta t} \Delta z \,, \tag{9.3}$$

where C and Δz are, respectively, the heat capacity and thickness of the soil layer.

For a homogeneous soil column, that is, one whose thermal conductivity does not change with depth, it is possible to determine from the temperature profile alone whether the different soil layers are warming or cooling. Several examples are shown in Figure 38, which is based on data obtained by Carson (1961) at the Argonne National Laboratory near Chicago, Illinois, during the summer of 1954. The direction and relative magnitude of the flux at selected depths are indicated by the arrows superimposed on the temperature curves. In each case, warming occurs if the net flux into a given layer is positive and cooling occurs if the net flux is negative.

The diurnal variation of the temperature profile into the soil follows a very regular course. The most prominent feature is a zone of high temperature, which originates at the surface and is the result of intense daytime heating by solar radiation. Beginning in the late afternoon, this zone propagates downward with decreasing intensity to deeper soil layers. As it moves, it loses heat by conduction to the cooler soil above and below. In Figure 38, the warm zone is located at depths of 5 to 40 cm in the late evening, 14 to 15 cm near sunrise, 45 to 80 cm at midday, and 50 to 85 cm in the late afternoon. By late afternoon it has almost disappeared.

The distribution of heating and cooling shown in Figure 38 could be changed considerably if the thermal conductivity varies with depth, as is practically always the case. For example, if the thermal conductivity increases with depth, the soil layer between the surface and 20 cm at midday and that between 20 and 50 cm in the late afternoon could both be regions of cooling, rather than of little change and warming, respectively, as indicated in the figure. Only when the direction of the flux changes through a layer is the sign of the temperature change certain.

Equation (9.1) is used to determine the flux of heat into or out of the soil, if the thermal conductivity and the temperature gradient at the soil surface are known. The thermal conductivity can be measured very accurately, in spite of its large variation with soil type and moisture content, with any one of several small heat-flow probes designed by De Vries and Peck (1957), Buettner (1955), Lachenbruch (1957), or Blackwell (1954), among others. These probes permit *in situ* measurements taken as frequently as desired and give more reliable results than the old method of measuring the heat flow through a soil sample placed on a plate and subjected to a constant heat source.

The temperature gradient at the soil surface is very difficult to determine, because

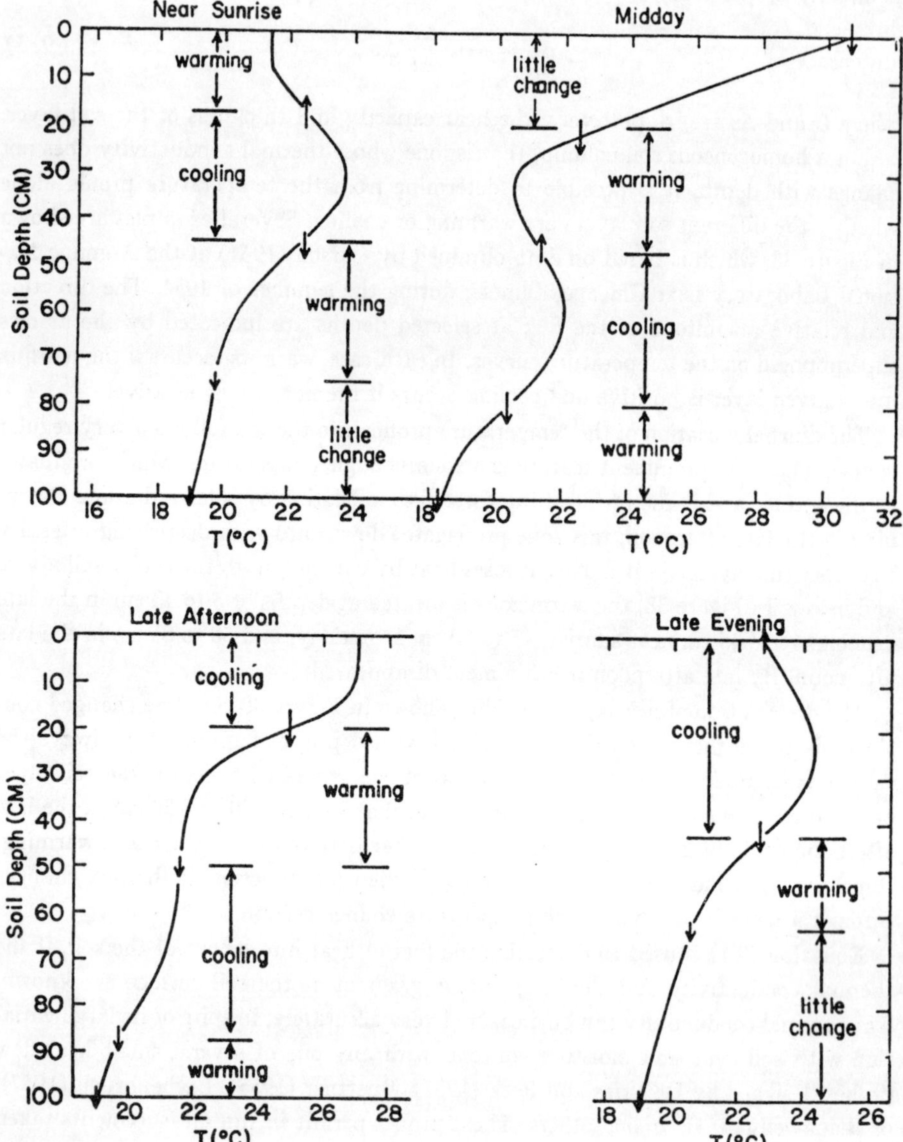

FIG. 38.—Typical variation of temperature with depth at different times of day in summer. Based on data given by Carson (1961) for a yellow-gray silt loam soil near Chicago, Illinois.

measurement of the surface temperature is necessary. A very small sensing element is required, since the temperature can change very rapidly with depth, especially in the middle of a warm summer day. Sinclair (1922), for example, measured a temperature difference of 21.1°C between 0.4 and 2.0 cm in soil at Tucson, Arizona, at 1300 on June 21, 1915.

To avoid this problem, small thin plates of known thermal conductivity with thermopile junctions embedded in the upper and lower faces are often used. For surface heat-flow measurements, these plates are placed not more than 2 cm below the soil surface. To take into account soil inhomogeneities and to obtain results representative of a given area, it is best to use more than one plate. These sensors are rugged and mobile, but they interfere with natural heat and moisture flow. Also, large errors may arise if the thermal conductivity of the plates differs significantly from that of the soil.

Perhaps the most accurate and most practical method for determining the flux of heat G into or out of the soil uses equation (9.3), from which

$$G = G_2 + C \frac{\Delta T}{\Delta t} z_2$$

where G_2 is the flux through a level z_2 5 to 10 cm below the surface and $\Delta T/\Delta t$ is the time rate of temperature change averaged for the soil layer from the surface to z_2. The former can be measured with a heat-flow plate and the latter with appropriate temperature sensors.

The heat capacity C that appears in equation (9.3) is much more regular and predictable than the thermal conductivity. According to De Vries (1958, 1963),

$$C = x_s C_s + x_w C_w + x_a C_a \tag{9.4}$$

where

$$C_s = \frac{x_m C_m + x_o C_o}{x_s}. \tag{9.5}$$

x_s, x_w, x_a, x_m, and x_o are, respectively, the volume fractions of solid matter, water, air, minerals, and organic matter in the soil, and C_s, C_w, C_a, C_m, and C_o are the corresponding heat capacities. According to De Vries (1963), C_m and C_o have average values of 0.46 and 0.60 cal cm^{-3} °C^{-1}, respectively, at 10°C. Then, since the heat capacity of air (0.0003 cal cm^{-3} °C) is much less than that of water (1.0 cal cm^{-3} °C) or of mineral and organic matter, equations (9.4) and (9.5) can be combined to give

$$C = x_s C_s + x_w \text{ cal cm}^{-3}\ {}^\circ\text{C}^{-1} \tag{9.6}$$

or

$$C = 0.46\,x_m + 0.60\,x_o + x_w \text{ cal cm}^{-3}\,°\text{C}^{-1}\,. \tag{9.7}$$

The heat capacity for a frozen soil will be less than that for an unfrozen soil, since $C_i = 0.46$ cal cm^{-3} °C^{-1} at 0 °C. In this case, since $x_i = 1.09\,x_w$, equation (9.6) becomes

$$C = x_s C_s + 0.50\,x_w \text{ cal cm}^{-3}\,°\text{C}^{-1}\,. \tag{9.8}$$

The following table shows the dependence of the heat capacity on the moisture content, expressed by the ratio x_w/x_{ad}, where x_{ad} is the volume fraction of air in the oven-dry soil. The data are taken from Vitkevich (1963). Except for peat, the heat

Soil Type	x_s	C_s	x_w/x_{ad}				
			0	0.2	0.5	0.8	1.0
Sand	0.585	0.517	0.302	0.385	0.510	0.634	0.717
Clay	.417	.576	.240	.357	.532	.706	.823
Peat	0.246	0.602	0.148	0.300	0.525	0.751	0.902

capacity is much less variable than the thermal conductivity. Its range is further restricted by the moisture contents likely to be encountered in the field.

In the absence of a water table, the moisture content of most soils normally lies between field capacity and the permanent wilting point. A soil is said to be at field capacity when it contains the maximum amount of water it can hold against the force of gravity. This is a vaguely defined term, since the drainage of water through soil does not cease abruptly at a fixed moisture content but rather proceeds at a steadily decreasing rate as the soil dries. For convenience, field capacity is frequently equated with a moisture tension or negative pressure exerted by the soil particles on the available water of one-third atmosphere (338 mb) for medium textured soils and one-tenth atmosphere for sandy soils (Richards and Richards, 1957).

As the soil dries, the moisture tension increases and it becomes more and more difficult for plants to extract and use the available water. Eventually, at a tension of about 15 atmospheres, the permanent wilting point is reached.

Representative values of the ratio x_w/x_{ad} at the wilting point and at field capacity are listed in Table 19. These data are not exact and are presented only to give the reader an idea of the range of values commonly observed. They are derived, for the most part, from papers by Livingston (1910), Russell (1939), Moore (1939), Lehane and Staple (1953), Richards and Richards (1957), Taylor (1957), De Vries (1958), and Burgos and Tschapek (1958).

The ability of a given soil to retain water, that is, its moisture tension, increases as the average pore size decreases and as the surface area per unit mass of soil, the specific surface, increases. The first factor is most important at high moisture contents and the second at low moisture contents. A sandy soil characteristically has large pores and a relatively low specific surface. Hence, the amount of water it can hold is limited, both at the wilting point and at field capacity. The moisture available for plant growth is usually small, and frequent watering or irrigation is necessary.

TABLE 19

REPRESENTATIVE MOISTURE CONTENTS AT THE WILTING POINT AND AT FIELD CAPACITY FOR THIRTEEN SOIL TYPES

(Values are expressed in terms of the volume fraction of air, x_{ad}, in the oven-dry soil)

SOIL TYPE	x_w/x_{ad}	
	Wilting Point	Field Capacity
Gravel	0.04	0.11
Coarse sand	.06	0.13
Medium sand	.10	0.26
Fine sand	.15	0.35
Sandy loam	.19	0.43
Fine sandy loam	.25	0.56
Silt loam	.30	0.66
Silty clay loam	.36	0.72
Clay loam	.32	0.74
Silty clay	.47	0.93
Clay	.48	0.93
Heavy clay	.65	1.21
Peat	0.21	0.62

A clay soil, on the other hand, is normally favored with very small pores and a remarkably high specific surface, of the order of seven times greater than that of sand (Richards and Richards, 1957). The total amount of usable water for plant growth between field capacity and the wilting point is much greater for clay than for other soils.

Heavy clay has the curious property of swelling as the moisture content increases; therefore, the volume fraction of water in the soil at field capacity actually exceeds the volume fraction of air in the oven-dry soil. The cracks formed when the soil dries out again show that swelling has occurred.

HEAT TRANSFER IN A HOMOGENEOUS SOIL

For an infinitely thin soil layer, equations (9.2) and (9.3) can be written in differential form and equated. Thus,

$$\frac{\partial G}{\partial z} = -\frac{\partial}{\partial z}\left(\lambda \frac{\partial T}{\partial z}\right) = -C \frac{\partial T}{\partial t}.$$

Further, if the soil is homogeneous, so that the thermal conductivity does not vary with depth, it follows that

$$\frac{\partial T}{\partial t} = \frac{\lambda}{C} \frac{\partial^2 T}{\partial z^2} = \kappa \frac{\partial^2 T}{\partial z^2}, \tag{9.9}$$

where $\kappa = \lambda/C$ is defined as the thermal diffusivity. This equation is the one-dimensional heat conduction equation and relates the warming or cooling of a soil column to the curvature of the temperature profile. Referring to Figure 38, we see that warming occurs wherever the curvature is positive and cooling wherever the curvature is negative.

The thermal diffusivities of various soil constituents are given in the following table. Values are also listed for stirred water and air under varying degrees of thermal stability (Priestley, 1959). For these the thermal diffusivity is greatly enhanced by the turbulence of the motion.

Substance	T(°C)	κ(cm² sec⁻¹)
Quartz	10	0.044
Clay minerals	10	0.015
Organic matter	10	0.0010
Ice	0	0.0114
Still water	10	0.0014
Stirred water		
Very stable		0.1
Moderately stable		50
Neutral		300
Still air	10	0.202
Stirred air		
Very stable		1.0×10^3
Neutral		1.0×10^5
Very unstable		1.0×10^7

The dependence of κ on x_w is shown in Figure 39 for the same soils that were used in Figure 37. For many soils the thermal diffusivity, which determines the heating or cooling rate accompanying a given temperature profile, is greatest at moisture contents of between 8 and 20 percent by volume. Within this range the thermal conductivity is not far removed from its maximum value and the heat capacity is relatively small. As a result, the soil conducts heat efficiently and warms or cools readily for a given energy input or output. The heating or cooling rate is impeded at very low moisture contents by the poor conductivity of the soil, the soil particles being well insulated by the air which fills the pore spaces, and at very high moisture contents by the large heat capacity of the soil, which permits only a modest temperature change per unit of heat assimilated.

Equation (9.9) can be solved for the simple case of a homogeneous semi-infinite soil whose surface is heated in a periodic manner that corresponds to the daily or

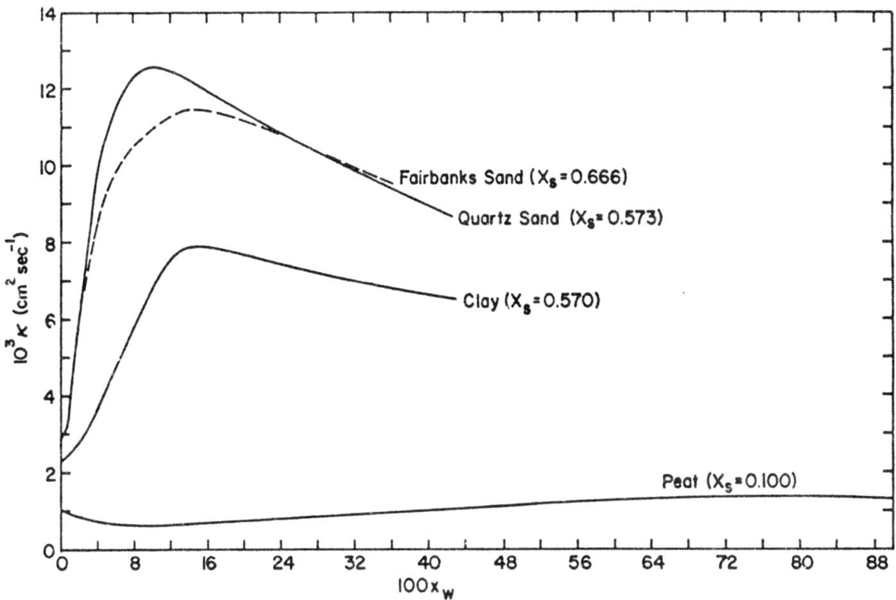

FIG. 39.—Dependence of the thermal diffusivity κ on the volume fraction of water x_w for four different soil types.

annual heating cycle experienced by the soil. It is assumed that the surface temperature, $T(0, t)$, at time t is given by

$$T(0, t) = \bar{T} + \Delta T_0 \sin \omega t \tag{9.10}$$

where $\bar{T}$ is the mean (daily or annual) soil temperature (assumed to be the same at all depths), ΔT_o is the amplitude of the surface temperature wave (half of the difference between the maximum and minimum surface temperatures), ω is the angular frequency of oscillation and equals $2\pi/P$, where P is the period of the wave (24 hours or

12 months). The time t is equal to 6 hours or 3 months at the time of maximum temperature.

With the above boundary condition, Carslaw and Jaeger (1959) show that the solution to equation (9.9) has the form

$$T(z, t) = \bar{T} + \Delta T_0 e^{-z(\omega/2\kappa)^{1/2}} \sin\left[\omega t - \left(\frac{\omega}{2\kappa}\right)^{1/2} z\right] \qquad (9.11)$$

where $T(z, t)$ is the temperature at the depth z at the time t. The quantity

$$\delta = \Delta T_0 e^{-z(\omega/2\kappa)^{1/2}}$$

represents the amplitude of the temperature wave at the depth z. Taking the ratio of the amplitudes at two depths, z_1 and z_2, we get the amplitude equation,

$$\frac{\delta_2}{\delta_1} = e^{-(z_2 - z_1)(\omega/2\kappa)^{1/2}}. \qquad (9.12)$$

Van Wijk and De Vries (1963) refer to the quantity $(2\kappa/\omega)^{1/2}$ as the damping depth D, since the amplitude is reduced to $e^{-1} = 0.37$ times its surface value at the depth $z = D$. The amplitude equals 0.01 ΔT_0 at $z = 4.61\ D$. Values of D and 4.61 D are listed below for different values of κ and for the daily and annual temperature cycles, for which $\omega^{1/2}$ equals 8.53×10^{-3} and 4.46×10^{-4} (rad sec^{-1})$^{1/2}$, respectively. The

κ(cm^2 sec^{-1})	Daily Cycle		Annual Cycle	
	D(cm)	4.61 D(cm)	D(m)	4.61 D(m)
0.001	5.2	24	1.0	4.6
0.004	10.5	48	2.0	9.2
0.008	14.8	68	2.8	13.1
0.012	18.2	84	3.5	16.0
15	6.4×10^2	30×10^2	123	566

thermal diffusivity of most soils lies between 0.001 and 0.012 cm^2 sec^{-1}. Hence, the daily temperature cycle penetrates to depths of 20 to 80 cm and the annual cycle to 5 to 20 m. On the other hand, in warm ocean currents, such as the Gulf Stream, the diffusivity averages about 15 cm^2 sec^{-1} and the daily and annual cycles penetrate to depths of 30 and 600 m, respectively.

From equation (9.11), the time of maximum soil temperature at any depth occurs when

$$\sin\left[\omega t - \left(\frac{\omega}{2\kappa}\right)^{1/2} z\right] = 1$$

or when

$$\omega t - \left(\frac{\omega}{2\kappa}\right)^{1/2} z = \frac{\pi}{2}.$$

Solving for t, applying to two depths, z_1 and z_2, and subtracting, we get the phase equation,

$$\Delta t_{max} = t_{max\,(2)} - t_{max\,(1)} = (z_2 - z_1)\left(\frac{1}{2\omega\kappa}\right)^{1/2}. \qquad (9.13)$$

The time of maximum (or minimum) temperature is delayed with increasing depth in the soil. The following table gives, for different thermal diffusivities, the depth z at which the time lag between the surface and z equals 3 hours for the daily cycle and 2 months for the annual cycle.

κ(cm² sec⁻¹)	Daily Cycle z(cm)	Annual Cycle z(m)
0.001........	4.1	1.0
0.004........	8.2	2.1
0.008........	11.7	2.9
0.012........	14.3	3.6
15............	5.0×10^2	127

The thermal diffusivity can be estimated from the amplitude and phase equations by solving for κ. Thus,

$$\kappa = \frac{\omega}{2}\left[\frac{z_2 - z_1}{\ln \delta_1/\delta_2}\right]^2 = \frac{1}{2\omega}\left[\frac{z_2 - z_1}{\Delta t_{max}}\right]^2 \qquad (9.14)$$

where natural logarithms are used. Both forms of equation (9.14) have been applied to natural soils by many investigators and have been found to give satisfactory results for the annual cycle, but not for the daily cycle. In general, the most consistent results are obtained with the phase relation. The failure of the theory is attributed in part to heat transfer by water (liquid and vapor) within the soil and in part to the heterogeneity of natural soils; the thermal conductivity typically changes with depth as well as with time as the moisture content varies. Both of these factors are most important in the upper meter or so of soil and, hence, have the greatest effect on the daily cycle. Lettau (1954), Van Wijk and De Vries (1963), and Van Wijk and Derksen (1963) have proposed more realistic theories of heat transfer in soil, taking into account time and depth variations of the thermal conductivity and the non-sinusoidal characteristics of the surface temperature wave.

In spite of its inadequacies, the simple theory outlined above provides considerable insight into the process of heat transfer occurring in the soil. When equation (9.1) is combined with equation (9.11) it follows that

$$G(z, t) = \Delta T_0(\omega C\lambda)^{1/2} e^{-z(\omega/2\kappa)^{1/2}} \sin\left[\omega t - \left(\frac{\omega}{2\kappa}\right)^{1/2} z + \frac{\pi}{4}\right]. \qquad (9.15)$$

In deriving this equation use has been made of the trigonometric identity

$$\sin y + \cos y = 2^{1/2} \sin\left(y + \frac{\pi}{4}\right).$$

At the surface, $z = 0$ and

$$G(0, t) = \Delta T_0 (\omega C\lambda)^{1/2} \sin\left(\omega t + \frac{\pi}{4}\right). \qquad (9.16)$$

Since $\pi/4$ corresponds to one-eighth of the period of oscillation, it follows by comparison with equation (9.10) that the time of maximum heat flux precedes the time of

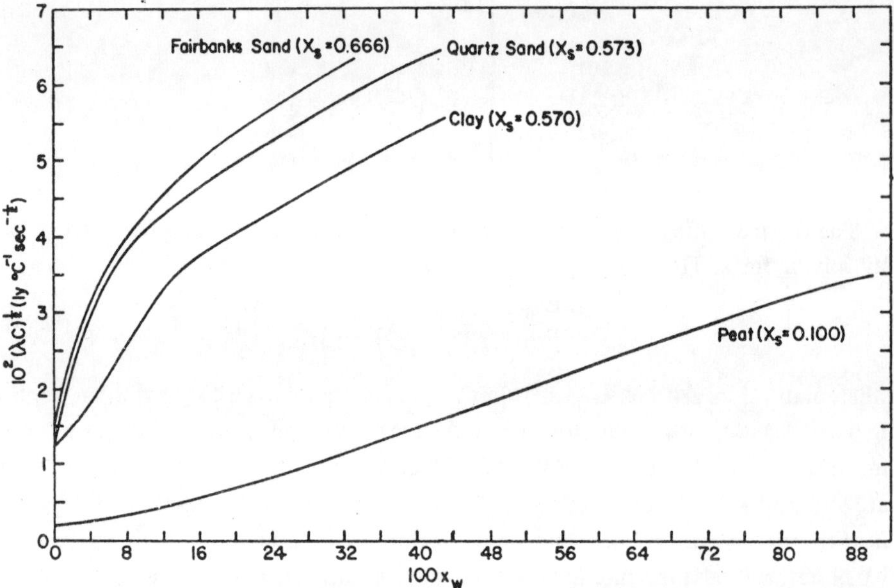

FIG. 40.—Dependence of the thermal property $(\lambda C)^{1/2}$ on the volume fraction of water x_w for four different soil types.

maximum surface temperature by 3 hours for the daily cycle and by one and a half months for the annual cycle. Equation (9.16) also predicts that the heat flux will drop to zero 3 hours or one and a half months after the time of maximum temperature. This behavior is well established by actual measurements.

The magnitude of the surface heat flux is proportional to the surface temperature amplitude ΔT_0 and to the quantity $(C\lambda)^{1/2}$ or $C\kappa^{1/2}$, which is called the conductive capacity by Priestley (1959) and Petterssen (1959), the thermal property by J. E. Johnson (1954), the soil product by Haltiner and Martin (1957), and the thermal contact coefficient by Businger and Buettner (1961) and Van Wijk and Derksen (1963). The term "thermal property" will be used here.

The amplitude ΔT_0 of the temperature wave at the soil surface is rarely known. De Vries (1958) relates it to the amplitude ΔT_2 of the air temperature at 2 m. He concludes that, for the annual cycle, the ratio $\Delta T_0/\Delta T_2$ over bare soil falls within the range of 1.1 to 2.0 and averages 1.3. Over short grass he assumes that $\Delta T_0 = 1.25\,\Delta T_2$. No similar relationships can be given for the daily cycle, because of the great variability involved.

The variation of the thermal property with moisture content is shown in Figure 40 for the four soils considered in Figures 37 and 39. In general, the range of values between the wilting point and field capacity is fairly small, from about 0.008 to 0.021 ly °C^{-1} sec$^{-1/2}$ for peat, from 0.033 to 0.042 ly °C^{-1} sec$^{-1/2}$ for sand, and from 0.044 to 0.056 ly °C^{-1} sec$^{-1/2}$ for clay.

The thermal property is of considerable importance in heat transfer problems, establishing the magnitude of the flux accompanying a given thermal regime. Equation (9.16) can be integrated from $t = -P/8$ to $t = 3P/8$ to give the total positive heat flux into the soil during the day or year. If we call this quantity $\Sigma G+$, it follows that

$$\Sigma G+ = 2\,\Delta T_0\,(\lambda C/\omega)^{1/2}\,. \tag{9.17}$$

Typical daily sums of the flux are listed below for different values of the thermal property and ΔT_0. The annual sums can be obtained by multiplying the figures given by $(365)^{1/2} = 19.1$.

	$\Sigma G+$ (ly)		
$(\lambda C)^{1/2}$ (ly °C^{-1} sec$^{-1/2}$)	0.01	0.03	0.05
$\Delta T_0 =$ 0°C........	0.0	0.0	0.0
2..........	4.7	14.1	23.4
4..........	9.4	28.1	46.9
6..........	14.1	42.2	70.3
8..........	18.8	56.3	93.8
10..........	23.4	70.3	117.2
15..........	35.2	105.5	176.0

According to Schmidt (1918), the partitioning r of energy between the soil and the air should be equal to the ratio of the thermal properties of the two media. That is

$$r = \frac{\Sigma G+}{\Sigma H+} = \frac{C\kappa^{1/2}}{\rho c_p K_h^{1/2}}$$

where ρ is the air density, c_p is the specific heat of air at constant pressure, and K_h is the thermal eddy diffusivity of air. In arriving at this conclusion, Schmidt assumed that all energy balance components vary periodically and that K_h is constant with height. The latter assumption is not valid. In the absence of free convection and with only forced convection acting, the eddy diffusivity should increase linearly with height. Both Lettau (1951) and Van Wijk and De Vries (1963) assume that

$$K_h = ku_* (z + z_o) \tag{9.18}$$

and arrive at the approximate solution

$$r = \frac{(\lambda C\omega)^{1/2}}{\rho c_p k u_*} \ln \frac{k u_*}{\gamma^2 z_o \omega} \tag{9.19}$$

where k and γ are constants equal to 0.4 and 1.781, respectively, and u_* and z_o are variables depending on the wind speed and the roughness of the surface, respectively. These will be discussed in more detail in chapter 10. For the present, it will suffice to say that u_*, called the friction velocity, is approximately equal to one-tenth of the wind speed at a height of 54.6 z_o cm, and that z_o, called the roughness length, is roughly equal to one-tenth the height of the vegetation or roughness elements.

Values of r, as a function of u_* and z_o, are listed below for a soil with a thermal property of 0.04 ly °C^{-1} sec$^{-1/2}$. Data are given for both the daily cycle and the annual

u_* (cm sec^{-1})	Daily Cycle z_o (cm)			Annual Cycle z_o (cm)		
	0.1	1.0	10.0	0.1	1.0	10.0
20.......	1.89	1.55	1.21	0.145	0.127	0.109
40.......	1.00	0.83	0.66	.075	.066	.057
60.......	0.68	0.57	0.46	0.051	0.045	0.039

cycle. The dependence of r on ω in equation (9.19) is attributed by Van Wijk and De Vries (1963) to the rapid increase of the eddy diffusivity with height. The longer the period of the oscillation, the better chance the surface heat wave has to penetrate to atmospheric levels where the convective heat transfer process is much more active and effective than it is close to the ground. As a result, the proportion of the energy entering the soil increases as the temperature fluctuations increase.

10/ TURBULENT TRANSFER AND WIND RELATIONSHIPS

TURBULENT TRANSFER PROCESSES

From observations of fluid flow through pipes, Reynolds (1883) distinguished between two types of motion, laminar motion and turbulent motion. He introduced some dye into the middle of a liquid current flowing through a thin glass tube and observed its behavior. When the flow velocity was low, a well-defined dye filament was visible, the fluid layers appearing to follow parallel courses. When the flow velocity was increased, however, the filament was torn apart and the dye distributed throughout the whole fluid, whose motion had become very irregular. The first type of flow is called laminar, the second, turbulent.

In laminar flow, the properties of a fluid can be transferred cross-stream only by the random heat motion of the fluid molecules. In turbulent flow, small blobs or eddies of the fluid break away from the mainstream and, after moving a short distance, mix with their new environment. These eddies serve as a very effective mechanism for transferring the physical properties of the fluid, that is, heat, matter, and momentum. In laminar flow through a pipe, the velocity decreases uniformly from a maximum at the center of the pipe to zero at the walls. In turbulent flow, most of the velocity decrease occurs near the walls, the bulk of the fluid moving at a uniform velocity. The fast-moving eddies at the center of the pipe tend to carry their momentum with them when they break away from the mainstream and transfer it to the slower moving fluid near the walls. This fluid, then, speeds up.

Atmospheric motions near the earth's surface are practically always turbulent, even when the air is very stable. This turbulent motion, however, cannot extend to the ground. A layer of air, at most a few millimeters thick, and usually much less than that, adheres with great tenacity to the surface. This is called the laminar boundary layer. Within this layer, heat, matter (water vapor), and momentum are transferred vertically only by molecular processes, and the rates of transfer are proportional to the vertical gradients of (potential) temperature T, specific humidity q, and horizontal air velocity u, respectively. Hence,

$$H = -\rho c_p \kappa_h \left(\frac{\Delta T}{\Delta z}\right)_o, \quad LE = -\rho L d \left(\frac{\Delta q}{\Delta z}\right)_o, \quad \text{and} \quad \tau = \rho \nu \left(\frac{\Delta u}{\Delta z}\right)_o \tag{10.1}$$

where τ represents the rate at which horizontal momentum of the air is being transferred vertically to the surface, and κ_h, d, and ν are, respectively, the thermal diffusivity of air (0.16 to 0.24 cm^2 sec^{-1}), the diffusivity of water vapor in air (0.20 to 0.29 cm^2 sec^{-1}), and the kinematic viscosity (0.12 to 0.17 cm^2 sec^{-1}).

Since the vertical heat flux from the ground can reach about 0.8 ly min^{-1} over a hot desert surface in the middle of a summer day, equation (10.1) implies that the temperature gradient within the laminar boundary layer can go as high as 30°C mm^{-1}.

If the surface is rough, to the extent that irregularities of the order of a few millimeters are present on it, a continuous laminar boundary layer will not exist. At the air-soil interface, however, molecular conduction or diffusion must always be the only non-radiative transfer process acting. Turbulent mixing cannot extend to the ground, as the following considerations will show.

As indicated in chapter 8, the vertical flux of sensible heat is virtually the same at 1 m as it is at the surface, if advective effects can be ignored. The same is true of the vertical flux of water vapor LE or momentum τ. Otherwise, unrealistic accelerations or changes of moisture content would occur within the layer. Hence, we can measure these quantities at 1 m, and sometimes even higher, and be fairly confident that the results give the surface fluxes.

Atmospheric turbulence near the ground is manifested by a rapid, and sometimes violent, variation of wind speed and direction superimposed on a mean flow, which remains steady for periods of 5 to 10 min or longer. Each lull or gust of wind speed recorded by a sensitive anemometer at 1 m represents the passage of an eddy. Since the mean wind speed usually increases with height, an eddy coming down from above will normally be traveling faster horizontally than the mean wind at the anemometer level and be sensed as a gust, and an eddy coming up from below will be traveling slower horizontally and be sensed as a lull.

Over a period of time, just about as much air passes upward through the 1-m level as passes downward, so that the mean vertical velocity is close to zero. (There may be exceptions to this, for example, strongly heated "hot spots.") There is, however, a net downward transfer of horizontal momentum, because the eddies moving down have more horizontal momentum than those moving up. Thus,

$$\overline{\rho w} \simeq 0\,, \quad \text{but} \quad \tau = -\overline{\rho w u} > 0\,. \tag{10.2}$$

Here, $\overline{\rho w}$ is the mean vertical mass transfer through the 1-m level per unit area and time, w is the vertical eddy velocity (positive upward), and u is the horizontal velocity or the horizontal momentum per unit mass. The units of τ are g cm^{-1} sec^{-2} or dynes cm^{-2}; τ actually represents a pressure force or shearing stress acting on the earth's surface.

The same type of analysis can be carried out for the vertical transfer of sensible heat and water vapor by turbulence. If the temperature decreases with height, the sensible heat content c_pT of a unit mass of air moving upward will be greater than that of a unit mass of air moving downward. Hence, the net sensible heat flux by turbulence through the 1-m level will be directed upward (or positive) without any corresponding mass flow, and

$$H = \rho c_p \overline{wT} \tag{10.3}$$

where the air density ρ is assumed to be constant. Similarly, for the vertical turbulent transport of water vapor or latent heat,

$$LE = \rho L \overline{wq}\,. \tag{10.4}$$

Equations (10.2), (10.3), and (10.4) implicitly assume that the rates at which momentum, heat, and water vapor are being transported vertically through the 1-m level by turbulence are equal to the rates at which the same three quantities are being transferred to or from the earth's surface by molecular processes. Turbulence cannot extend to the ground, because the vertical velocity vanishes there. Its importance, however, increases rapidly with height in the lowest few centimeters. On the other hand, molecular conduction is most important very close to the ground, where the vertical gradients of temperature, specific humidity, and wind speed are greatest, and insignificant above 1 or 2 mm. Obviously, there must be some thin air layer close to the ground where both processes are of equal importance.

In the absence of turbulence, the diurnal temperature variation at 1 m would be only about a quarter of that observed at the surface. Further, the maximum temperature at 1 m would occur 5 hours after the surface maximum.

Theoretically, it would be possible to obtain the vertical fluxes of heat, water vapor, and momentum directly by measuring the vertical gradients of temperature, specific humidity, and wind speed in the laminar boundary layer. But this is really not a very promising approach, except possibly over a flat plate in the laboratory or over a smooth ice sheet, because the laminar boundary layer is so thin and ill defined. Vehrencamp (1951), measuring the air temperature at eight levels below 25 mm (four levels below 1.5 mm), attempted to determine H by this method but had only partial success.

More promising are the turbulent transfer equations

$$\tau = -\rho\overline{wu}\,, \quad H = \rho c_p\overline{wT}\,, \quad \text{and} \quad LE = \rho L\overline{wq}$$

which make up the basic relations in the eddy correlation method for determining the vertical fluxes of heat, water vapor, and momentum. They have been applied with moderate success, especially by Swinbank (1955) and Dyer (1961) in Australia and by Frankenberger (1960) in Germany, but it is very doubtful that they will ever come into general use. Extremely fast responding, well-matched instruments are required in order to detect the smallest eddies contributing to the flux. Better results are achieved by going to higher levels, for example, 5 or 6 m, where the important eddies are larger and hence easier to detect. But then storage and advective effects below the measuring level may produce serious errors. (See, for example, Dyer and Pruitt, 1962.)

The most common instrumental methods for measuring the vertical fluxes are based on the assumption that the same type of equation that governs molecular transfer also applies to turbulent transfer, if the molecular diffusivities, κ_h, d, and ν are replaced by their corresponding eddy diffusivities, K_h, K_w, and K_m, respectively. Thus,

$$H = -\rho c_p K_h \frac{\Delta T}{\Delta z}, \tag{10.5}$$

$$LE = -\rho L K_w \frac{\Delta q}{\Delta z}, \tag{10.6}$$

and

$$\tau = \rho K_m \frac{\Delta u}{\Delta z}. \tag{10.7}$$

Here the gradients are normally measured between 0.5 and 2 m above the surface, values averaged over periods of 30 to 60 min being used.

When applying the gradient equations, a problem arises in evaluating the exchange coefficients, which unlike their molecular counterparts, are tenuous quantities subject to large variations with both height and time of day or atmospheric stability. In general, at 1 m their values will range from about 10^2 cm^2 sec^{-1} at night when the temperature inversion is strongest to 10^5 cm^2 sec^{-1} in the early afternoon when heating of the surface air layer is most intense. They increase almost linearly with height, at least within a few tens of meters of the ground.

Alternate expressions for H, LE, and τ can be obtained if we assume that the total flux does not vary with height. For sensible heat, for example, summing equation (10.5) and the first of equations (10.1) and integrating from the surface to a height, z, 1 or 2 m above the ground gives

$$H = \rho c_p D_h (T_s - T). \tag{10.8}$$

Also, $$LE = \rho L D_w (q_s - q) \quad \text{or} \quad LE = 0.622 \frac{\rho L D_w}{p} (e_s - e) \tag{10.9}$$

and $$\tau = \rho D_m u \tag{10.10}$$

where, for example,

$$D_h = \left(\int_0^z \frac{1}{K_h + \kappa_h} dz \right)^{-1}. \tag{10.11}$$

T_s, q_s, and e_s are the surface temperature, specific humidity, and vapor pressure, and T, q, e, p, and u are the air temperature, specific humidity, vapor pressure, total pressure, and wind speed, respectively.

In several respects, equations (10.8) to (10.10) are easier to work with than equations (10.5) to (10.7). The air temperature, humidity, and wind speed need be measured at only one level. Also, the transfer coefficients, D_h, D_w, and D_m, vary only slightly with height above 1 m. Drawbacks include the difficulties involved in measuring the surface temperature and vapor pressure.

To avoid evaluating them, we frequently assume that the eddy diffusivities are equal. This is not a bad assumption in the lowest meter of air, where most turbulence is associated with the natural roughness of the terrain, but at greater heights buoyancy forces become important and act selectively to transfer heat upward more readily than either momentum or water vapor.

If $K_h = K_w = K_m$, it follows that

$$\frac{H}{LE} = \frac{c_p}{L} \frac{\Delta T}{\Delta q}, \quad \frac{\tau}{LE} = \frac{-\Delta u}{L \Delta q}, \quad \text{and} \quad \frac{\tau}{H} = \frac{-\Delta u}{c_p \Delta T}$$

where $\Delta T = T_2 - T_1$, $\Delta q = q_2 - q_1$, and $\Delta u = u_2 - u_1$. These ratios are easy to measure, but they still do not give the actual magnitudes of the fluxes. This can be done by first obtaining either H, LE, or τ by another method. Sometimes a lysimeter is available for direct measurement of the evaporation rate, and, hence, of LE (see chap. 11), but more often the sensible heat flux is eliminated from the first ratio with the energy balance equation for the earth's surface. Solving for LE, we get

$$LE = \frac{R - G}{1 + \frac{c_p}{L} \frac{\Delta T}{\Delta q}} \tag{10.12}$$

and

$$H = R - G - LE.$$

If R, G, ΔT, and Δq are measured, the latent heat flux can be obtained; the sensible heat flux is then the residual in the energy balance equation. This method has been used extensively in the United States by Harbeck *et al.* (1958) and by Tanner (1963) and his associates. It breaks down, however, when R is equal to G and when the ratio H/LE is close to minus one, as is frequently the case with an inversion at night.

To make practical use of the ratio of the momentum flux to the latent or sensible heat flux, more must be known about the variation of the wind speed with height near the ground. It will be shown in the next section of this chapter that below 1 m under average conditions

$$\tau = \rho k^2 \left(\frac{\Delta u}{\ln z_2/z_1}\right)^2 \tag{10.13}$$

where k is a constant approximately equal to 0.4 and $\ln z_2/z_1$ is the natural logarithm (base e) of the ratio of the heights, z_2 and z_1, at which the wind speed is measured. It then follows that

$$LE = -\rho L k^2 \frac{\Delta u \Delta q}{(\ln z_2/z_1)^2} \tag{10.14}$$

and

$$H = -\rho c_p k^2 \frac{\Delta u \Delta T}{(\ln z_2/z_1)^2}. \tag{10.15}$$

These are the basic equations in the aerodynamic method for determining the vertical fluxes of heat, water vapor, and momentum. They have been widely used, especially over water surfaces, and apparently give good results, at least when the vertical temperature gradient is not too steep. The method has been used successfully over land by House, Rider, and Tugwell (1960).

The eddy correlation, energy balance, and aerodynamic methods, together with lysimeter measurements of evaporation, are the generally accepted methods for measuring the fluxes of latent and sensible heat from the ground to the atmosphere. However, none is completely accurate and all involve assumptions, which under certain conditions may not be valid. The most serious assumption common to all methods is the neglect of the horizontal advection of heat and moisture below the measuring level. For this to be a reasonable assumption, measurements should be made below 1 m in the middle of large fields or plains where the horizontal gradients of temperature and moisture are small. This is not always possible where the terrain is broken and clearings are interspersed with brush or trees.

The size of the uniform area required to give representative results has been the subject of several observational and theoretical studies. It depends, to a large extent, on the depth of the air layer being sampled. The following table, derived from data taken by Rider, Philip, and Bradley (1963), shows the dependence of evaporation rates from short grass, as computed using equation (10.12), on the distance x, downwind from a large, dry tarmac surface and on the heights, z_1 and z_2, above the ground at which temperatures and specific humidities are measured. The observations were taken at Canberra, Australia, during January and February of 1961. The values given in the table are based on data for sixteen 10-min observation periods between 1300 and 1600 local time. The net radiation ranged from 0.720 to 0.882 ly min^{-1} and the wind speed at 2 m averaged 512 cm sec^{-1}.

		E (mm hour^{-1})			
z_1(cm)	z_2(cm)	$x = 0$ m	$x = 1.1$ m	$x = 4.3$ m	$x = 17.3$ m
5	11.5	−0.318	0.535	0.758	0.818
11.5	27.5	− .079	.175	.634	.793
27.5	64	− .016	− .039	.291	.731
64	150	−0.063	−0.296	0.033	0.545

Since the grass was watered twice a week, it is reasonable to assume that evaporation proceeded at close to the potential rate, which, for a very large grass area, nearly equals $(R - G)L^{-1}$, or 0.751 mm hour^{-1} for these data. The actual evaporation rate at 17.3 m downwind has been estimated by the author, and also by Rider *et al.* (1963), to be of the order of 1.0 mm hour^{-1}. It follows, then, from the above table that meaningful results are obtained only when the ratio of the upwind fetch distance to the height of the upper sensor is greater than about 25. The same ratio applies to the observed wind profile data and the momentum flux. No explanation is given for the negative evaporation rates obtained from the temperature and moisture gradients near the tarmac-grass boundary.

Theoretical estimates of the fetch-height ratio range from 10 (Panofsky and Townsend, 1964) to 140 or more (Dyer, 1963). Much of the disagreement is simply a matter of definition. The large values obtained by Dyer (1963), for example, refer to the downward distance at which the actual flux at a given height is within 10 percent of that at the same height in the center of an infinitely large field.

Sellers and Hodges (1962) experimented with a method that would give the fluxes of sensible and latent heat from very small areas. The method essentially involves measuring the sensible heat and moisture picked up by air drawn through a small plastic-covered tunnel placed on the ground. The dimensions of the tunnel (about 10 cm high, 30 cm wide, and 50 cm long) and the air flow rates (about 1 m sec^{-1}) are such that most sensible heat sent out from the ground is transported horizontally out of the tunnel, and only a small fraction is lost by conduction through the plastic. Thin (0.5-mil) polyethylene is used for the plastic covering because this material permits maximum transmission of radiation at all wavelengths. It is obvious that in forcibly drawing air through the tunnel the natural environment is modified. As long as the air flow rate is reasonable, however, this does not seem to be a serious problem. The method is recommended only over bare soil or very short grass. Attempts to use it to estimate transpiration from shrubs or trees will, in general, not be successful.

WIND RELATIONSHIPS

The characteristics of the wind and temperature profiles close to the ground can best be explained by referring to a typical example. In Figure 41 the average variation of temperature (upper part) and wind speed (lower part) with height below 16 m is shown for four different hours (0600, 1400, 1900, and 2200 CST). The curves are derived from observations taken over short prairie grass by Covey *et al.* (1958) at O'Neill, Nebraska, in the summer of 1956. Both quantities were measured at heights of 0.25. 0.50, 1, 2, 4, 8, and 16 m above the ground; in addition, the temperature was also measured at 0.125 m. A logarithmic scale is used for the height. The observed profiles are given by the solid lines; the dashed lines are the best-fitting straight lines through the three lowest observed points.

At 0600 and 1900, neutral stability conditions prevail and the temperature is nearly constant with height. At these times the wind speed increases almost linearly with the logarithm of the height, at least below 8 m. As a result, the rate of change of wind speed with the logarithm of the height is constant. That is

$$\frac{u_2 - u_1}{\ln z_2 - \ln z_1} = \frac{\Delta u}{\Delta(\ln z)} = \frac{\Delta u}{\ln z_2/z_1} = \text{constant}.$$

If it is assumed that the wind speed vanishes at some non-zero height z_o then it follows that

$$u = (\text{constant}) \ln z/z_o$$

where u is the wind speed at the height z.

The constant is the slope of the line, and depends partly on the rate at which momentum is transferred vertically τ and partly on the air density ρ. Laboratory experiments of turbulent flow through pipes show that

$$\text{Constant} = \frac{1}{k}\left(\frac{\tau}{\rho}\right)^{1/2} = \frac{u_*}{k}$$

where k is a constant of proportionality, approximately equal to 0.4 (the Von Karman constant). Hence,

$$\frac{\Delta u}{\ln z_2/z_1} = \frac{1}{k}\left(\frac{\tau}{\rho}\right)^{1/2} = \frac{u_*}{k} \tag{10.16}$$

and

$$u = \frac{u_*}{k} \ln \frac{z}{z_o}. \tag{10.17}$$

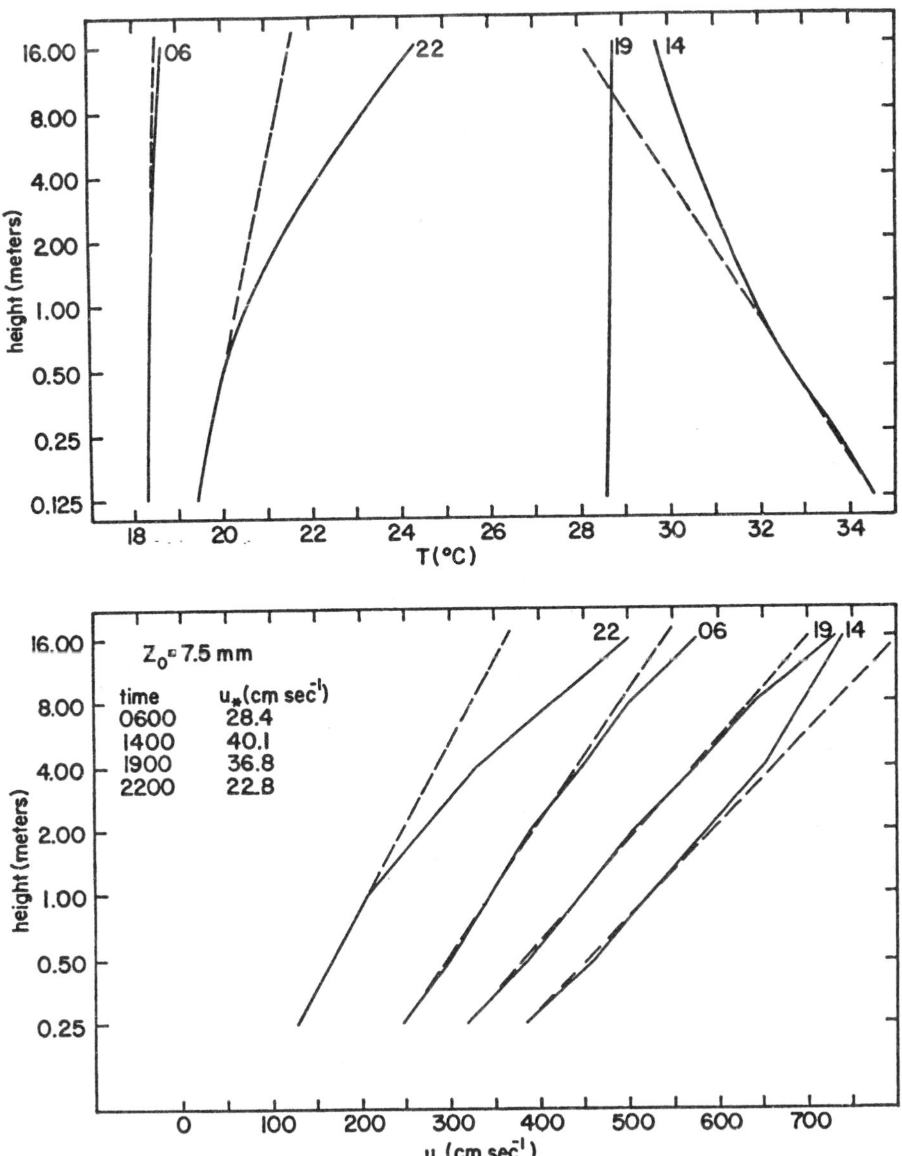

FIG. 41.—The average variation of temperature (*top*) and wind speed (*bottom*) with height and time of day in the air layer near the ground. The observations were taken over short grass at O'Neill, Nebraska, during the summer of 1956 by Covey *et al.* (1958).

Equation (10.13) follows directly from equation (10.16), when the latter is solved for τ.

The quantities u_* and z_o are called the friction velocity and roughness length, respectively. They can be obtained easily from the wind profile, since k/u_* is the slope of the line and z_o is its intercept with the z axis (where $u = 0$). For the data given in Figure 41, the friction velocity is equal to 28.4 cm sec^{-1} at 0600 and 36.8 cm sec^{-1} at 1900, or roughly one-tenth of the wind speed at 0.4 m. By extrapolation, the roughness length is found to be about 7.5 mm at both 0600 and 1900, which can reasonably be expected, since it is mainly a function of the type of surface, in this case, mowed prairie grass.

Typical roughness lengths for various surface types are listed in Table 20. For tall vegetation, equation (10.17) is frequently replaced by

$$u = \frac{u_*}{k} \ln \frac{z - d}{z_o} \tag{10.18}$$

where d is called the zero-plane displacement, since the wind speed goes to zero at the height $z_o + d$, rather than at z_o, as in equation (10.17). Observed values of d are so variable and random (occasionally being negative) that it is almost impossible to

TABLE 20

ROUGHNESS LENGTHS FOR VARIOUS SURFACES

Type of Surface	h(cm)	z_o(cm)	Author
Fir forest	555	283	Baumgartner (1956)
Citrus orchard	335	198	Kepner *et al.* (1942)
Large city (Tokyo)		165	Yamamoto and Shimanuki (1964)
Corn	300		Wright and Lemon (1962)
*$u_{5.2}$=35 cm sec^{-1}		127	
$u_{5.2}$=198 cm sec^{-1}		71.5	
Corn	220		Wright and Lemon (1962)
$u_{4.0}$=29 cm sec^{-1}		84.5	
$u_{4.0}$=212 cm sec^{-1}		74.2	
Wheat	60		Penman and Long (1960)
$u_{1.7}$=190 cm sec^{-1}		23.3	
$u_{1.7}$=384 cm sec^{-1}		22.0	
Grass	60–70		Deacon (1953)
$u_{2.0}$=148 cm sec^{-1}		15.4	
$u_{2.0}$=343 cm sec^{-1}		11.4	
$u_{2.0}$=622 cm sec^{-1}		8.0	
Alfalfa brome	15.2		Tanner and Pelton (1960*b*)
$u_{2.2}$=260 cm sec^{-1}		2.72	
$u_{2.2}$=625 cm sec^{-1}		2.45	
Grass	5–6	0.75	Figure 41
	4	0.14	Rider *et al.* (1963)
	2–3	0.32	Rider (1954)
Smooth desert		0.03	Deacon (1953)
Dry lake bed		0.003	Vehrencamp (1951)
Tarmac		0.002	Rider *et al.* (1963)
Smooth mud flats		0.001	Deacon (1953)

* The subscript gives the height (in meters) above the ground at which the wind speed u is measured.

attach any real physical significance to them. Further, in practically all cases and within the accuracy of the measurements, the wind profile above the vegetation can be described just as well by equation (10.17) as by equation (10.18). Therefore, roughness lengths obtained from profile data presented by Baumgartner (1956), Kepner *et al.* (1942), Wright and Lemon (1962), Penman and Long (1960), Deacon (1953) for grass, and Tanner and Pelton (1960*b*), each of whom used equation (10.18), have been recomputed using equation (10.17). The results are given in the table.

For natural vegetation that yields to the wind, the roughness length tends to decrease with increasing wind speed, presumably because at high speeds the vegetation presents a relatively flat and smooth surface to the air flow.

As noted by Tanner and Pelton (1960*a*), the relationship between the roughness length and the vegetation height h can be expressed quite adequately by

$$\log z_o = a + b \log h$$

where common (base 10) logarithms are used. From the data of Table 20, with z_o and h in centimeters, $a = -1.385$ and $b = 1.417$. These values compare favorably with those obtained by Tanner and Pelton (1960*a*), where $a = -0.883$ and $b = 0.997$, and by Kung (1961, 1963), where $a = -1.24$ and $b = 1.19$.

Equation (10.17), when combined with equations (10.7) and (10.10), gives the following values for the eddy viscosity K_m and the momentum transfer coefficient D_m:

$$K_m = k u_* z = k^2 u z \, (\ln z/z_o)^{-1} \qquad (10.19)^1$$

$$D_m = u_*^2/u = k^2 u \, (\ln z/z_o)^{-2} \, . \qquad (10.20)$$

The actual wind velocity goes to zero at the surface and not at the height z_o predicted by equation (10.17). This discrepancy is explained since, first, the logarithmic law does not apply within a vegetation canopy and, second, fully turbulent flow, which is required for the law to hold, does not occur in the lowest few millimeters over a smooth surface. Here, a significant part of the momentum transfer is by molecular processes acting under the influence of the strong wind shear that normally exists.

Nevertheless, in order to satisfy the boundary condition that $u = 0$ at $z = 0$, equation (10.17) is frequently written

$$u = \frac{u_*}{k} \ln \frac{z + z_0}{z_0} \, . \qquad (10.21)$$

This relationship yields non-zero values for the eddy viscosity at the surface, where turbulent motion is presumably absent. For this reason, it does not seem to be as valid as equation (10.17).

[1] To be more exact, u_*^2 should be defined as the ratio of the total (molecular plus turbulent) momentum transfer to the air density. Then the left-hand side of equation (10.19) becomes $K_m + \nu$. Except very close to the ground, $K_m \gg \nu$.

Under stable conditions, when the temperature increases with height (at time 2200 in Figure 41), and unstable conditions, when the temperature decreases with height (at time 1400 in Figure 41), equation (10.17) continues to hold at heights below 1 m. The temperature itself also varies linearly with the logarithm of the height below 0.5 m, the form of the relationship being

$$T_2 - T_1 = \Delta T = -\frac{H}{\rho c_p k u_*} \ln \frac{z_2}{z_1}, \tag{10.22}$$

which, with the aid of equation (10.16), reduces to equation (10.15). In Figure 41, the friction velocity is equal to 22.8 cm $\sec^{-1}$ at 2200 and 40.1 cm $\sec^{-1}$ at 1400, giving values for the sensible heat flux of -0.068 and 0.372 ly $\min^{-1}$, respectively.

Above 1 m in non-neutral conditions both the wind and temperature profiles depart from the logarithmic law in the sense that the gradients are less steep when the temperature decreases with height and steeper when the temperature increases with height. These results, which are commonly observed, can be easily explained. Under stable conditions there is relatively little vertical exchange. Turbulence is still present, but it is considerably weaker than under neutral conditions. As a result, the momentum of the fast-moving upper layers is not easily mixed with that of the slower-moving lower layers. Also, the radiative cooling of the surface air is not readily overcome by vertical convection.

Under unstable conditions, turbulence is increased. The natural buoyancy of the strongly heated air close to the ground causes it to rise and to be replaced by cooler air descending from above. Both heat and momentum are easily mixed in this manner, producing weaker vertical gradients of wind speed and temperature than exist under neutral conditions when buoyancy effects are absent and turbulent mixing or convection is mechanically induced by the terrain and the wind shear. In any case, mechanical turbulence usually dominates in the lowest meter.

There have been numerous attempts to derive a mathematical expression for the variation of wind with height under non-neutral conditions, but none has been completely satisfactory, and practically all fail under very stable conditions, when, according to Lettau (1962), there is a strong coupling between the air flow near the ground and that at greater heights.

Two empirical relations suggested by Hellmann (1919) and Deacon (1949) have been used extensively. According to Hellmann,

$$\frac{u_2}{u_1} = \left(\frac{z_2}{z_1}\right)^a$$

where a is a variable parameter depending on the stability of the air layer. On logarithmic paper this power law plots as a straight line with a slope of a. It fits the wind profiles in Figure 41 reasonably well if $a = 0.18$ at 0600 and 1900 (neutral), $a = 0.14$ at 1400 (unstable), and $a = 0.33$ at 2200 (stable). According to Sutton (1953), values of a given in the literature range from almost zero under very unstable conditions (Ali, 1932) to 0.85 under very stable conditions (Best, 1935). As pointed out by Molga (1962), a also decreases with height as frictional interaction with the earth's surface becomes less. For the data of Figure 41, the average 24-hour value of a decreases from 0.30 for the air layer between 0.25 and 0.50 m to 0.19 for the layer from 4 to 8 m.

The power law has been used in diffusion theory, because it is easier to handle mathematically than equation (10.17). It is also a convenient expression to extrapolate the wind speed at one level to that at another level.

Deacon (1949) proposed that, under general stability conditions, equation (10.16) can be written

$$\frac{\Delta u}{\ln z_2/z_1} = \frac{u_*}{k}\left(\frac{z}{z_o}\right)^{1-\beta}$$

where β is a stability parameter equal to one under neutral conditions, greater than one under unstable conditions, and less than one under stable conditions. The quantity $1 - \beta$ is the slope of the line when the logarithm of the left-hand member of the equation is plotted as a function of the logarithm of the height. Using observations taken at O'Neill, Nebraska, in the summer of 1953, Davidson and Barad (1956) obtained values of β ranging from 0.60 to 1.10. They noted that β decreases with height under stable conditions and increases with height under unstable conditions.

There have been numerous attempts to improve upon the power law and the Deacon profile. At least eight different "diabatic profiles" have appeared in the literature, each one supposedly fitting the observations. Most of these relate the variable

$$\varphi = \frac{k\Delta u}{u_* \ln z_2/z_1} = \frac{kz\Delta u}{u_*\Delta z} \tag{10.23}$$

which equals one under neutral conditions, to the Richardson number,

$$Ri = \left[\frac{g\Delta T}{T(\Delta u)^2}\ln\frac{z_2}{z_1}\right]z \tag{10.24}$$

where the quantity in brackets varies only slowly with height.

The Richardson number represents the ratio of the rate at which mechanical energy for the turbulent motion is being dissipated (or produced) by buoyancy forces (free or natural convection) to the rate at which mechanical energy is being produced by inertial forces (forced or mechanical convection). Turbulent energy is always increased by forced convection when air flows over a rough surface. Free convection, on the other hand, is an energy source only when the temperature decreases with height ($Ri < 0$) and an energy sink when the temperature increases with height ($Ri > 0$).

In the first case, an air parcel displaced upward will be less dense than its environment and will accelerate. In the second case, the parcel will be denser than its environment and will tend to return to its original level.

Thus, with a stable thermal stratification, turbulent motion is suppressed by buoyancy forces. According to Lumley and Panofsky (1964), atmospheric turbulence cannot be maintained at Richardson numbers greater than about 0.25. Such values occur fairly often on clear nights with strong inversions and light winds. At the other extreme, values of -0.70 or less are not uncommon on warm summer days.

As implied by equation (10.24), the Richardson number normally increases in absolute magnitude with height. Hence, forced convection near the ground is usually replaced by free convection at greater heights, at least under unstable conditions. Under stable conditions, the transition may be to a type of quasi-laminar flow.

Priestley (1955, 1959) considers the transition from forced to free convection under unstable conditions to be an abrupt one, which occurs at Richardson numbers between -0.02 and -0.05. Other investigators have suggested interpolation formulas that give a gradual transition from one regime to the other. As noted by Lettau (1962), the form of the relationship normally used is

$$\varphi = (1 - bRi)^{a} . \qquad (10.25)$$

From dimensional analysis of free and forced convection, Ellison (1957) finds that the coefficient a should equal -0.25, since in the presence of free convection alone

$$\varphi \propto (-Ri)^{-1/4} .$$

Hence,

$$\varphi = (1 - bRi)^{-1/4} . \qquad (10.26)$$

The coefficient b must be determined experimentally. Panofsky, Blackadar, and McVehil (1960) obtained a value of 18 from an analysis of data from several sources. This is in general agreement with values obtained by other investigators. Most of these fall within the range from 14 to 24. Equation (10.26) has been derived independently by Kazanski and Monin (1956), Ellison (1957), Yamamoto (1959), Panofsky (1961), and Sellers (1962). Using the initial letter of each one of its inventors' names, Panofsky (1963) calls it the "KEYPS" profile. Yamamoto and Shimanuki (1964) find that the profile holds to heights of 200 m over the city of Tokyo under extremely unstable and near-neutral conditions.

For small values of the product bRi, that is, under near-neutral conditions, equation (10.26) may be approximated by

$$\varphi = (1 + \tfrac{1}{2}bRi)^{1/2} \qquad (10.27)$$

$$\varphi = (1 - \tfrac{1}{2}bRi)^{-1/2} \qquad (10.28)$$

or

$$\varphi = (1 - \tfrac{1}{4}bRi)^{-1} . \qquad (10.29)$$

The first form has been used by Rossby and Montgomery (1935) and Sverdrup (1936), the second by Holzman (1943) and Budyko (1963*b*), and the third by Monin and Obukhov (1954), Elliott (1960), and McVehil (1964).

Of the forms presented here, only equation (10.29) can be integrated directly. As shown by Lumley and Panofsky (1964),

$$Ri = z/\varphi L'$$

where L' is a gradient length and is given by

$$L' = \frac{u_* T \Delta u}{k\, g \Delta T} .$$

Substituting this into equation (10.29) gives

$$\frac{\Delta u}{\Delta z} = \frac{u_*}{k}\left(\frac{1}{z} + \frac{b}{4L'}\right).$$

If L$'$ is constant with height (see Barad, 1963), it follows that

$$u = \frac{u_*}{k}\left[\ln \frac{z}{z_o} + \frac{b}{4L'}(z - z_o)\right]. \qquad (10.30)$$

Equation (10.30) is frequently called the "log-linear" law. McVehil (1964) found that it gives a good fit to observed wind data under stable conditions as long as the Richardson number is less than 0.15 and if b is 28.

More complex, but not necessarily more accurate, diabatic wind profiles have been suggested by Lettau (1952), Businger (1959), Webb (1960), and Swinbank (1964). All of these, including those mentioned above, are of diagnostic, rather than predictive, value, since the wind shear always appears on the right-hand side of the equations. Only the simple power law and the logarithmic law can be used directly to estimate the wind speed at one level from that at another. Neither, however, is applicable where the wind is being strongly accelerated, such as in the vicinity of tornadoes or dust devils.

11 / EVAPORATION AND EVAPOTRANSPIRATION

The rate at which water is lost from the surface, either by transpiration or evaporation, is of great importance in determining the water needs of any region. Survival in the desert, for example, is partly dependent on whether water can be supplied faster than it is lost by evaporation and transpiration, principally from reservoirs and large irrigated fields. The estimated annual evaporation from Lake Mead is 2.2 m, or about 4 percent of the capacity of the reservoir. In midsummer, between 25 and 40 cm of water will be lost by evaporation from small water bodies (ponds, stock tanks, swimming pools) in a single month.

The investigator has his choice of many methods for measuring or estimating evaporation rates and the latent heat flux. That so many exist indicates that none is completely adequate for all purposes. The methods can be separated into three groups: first, those used primarily in micrometeorological research and discussed in the previous chapter; second, those involving containers filled with water or soil (evaporimeters and lysimeters); and third, those based on climatological data and used to estimate long-period variations in evaporation.

EVAPORIMETERS AND LYSIMETERS

It is possible to distinguish between and discuss three different types of water- or soil-containing units for estimating evaporation or transpiration. These will be referred to as water evaporimeters, soil evaporimeters, and lysimeters.

Water evaporimeters.—By far the most common device of this type is a water-filled tank or pan. Evaporation is measured by noting the change in water level during a short time period (usually one day). In 1915, the United States Weather Bureau set up a network of stations around the country to measure evaporation of water from what is known as the Weather Bureau Class A land pan. The first station was at Roosevelt Reservoir in Arizona. The pan is 122 cm in diameter and 25 cm deep and is placed on a spaced-timber platform, so that its bottom is about 14 cm above the ground. A 3-cup totalizing anemometer, with the cups 61 cm above the ground, is normally mounted on the supports of the pan. There are many other types of evaporation pans, with many sizes, shapes, and exposures, but the Class A pan is by far the most common in the United States.

TABLE 21

SUMMARY OF PAN AND LAKE EVAPORATION DATA

LAKE	AREA (km²)	DEPTH (m)	PERIOD	EVAPORATING SURFACE	EVAPORATION RATE (mm day⁻¹)				ANNUAL TOTAL (mm)	SOURCE
					Winter	Spring	Summer	Fall		
Salton Sea (33.5°N)...	777	7.4	1961–62	Lake	2.25	7.03	7.94	3.26	1,873	Hughes (1963)
				Class A pan	5.74	14.07	13.96	5.94	3,627	
				Sunken pan	4.15	9.89	10.26	4.90	2,669	
Mead (36.2°N)...	514	53.6	1952–53	Lake	3.93	6.49	7.65	5.75	2,177	Harbeck (1958)
				Class A pan	4.90	11.64	11.72	3.91	2,939	Kohler *et al.* (1958)
				Floating pan	4.14	9.01	10.25	5.05	2,600	
Hefner (35.5°N)...	9	8.2	1950–51	Lake	1.78	3.72	5.46	4.23	1,390	Kohler (1954)
				Class A pan	3.76	7.66	7.12	3.98	2,057	
				Sunken pan	3.14	5.45	4.92	3.56	1,559	
Mendota (43.1°N)...	39	12.2	27 years	Lake	.....	1.53	4.44	2.89	814	Dutton and Bryson (1962)
				Class A pan[1]	0.53	4.79	4.43	1.09	991	
Towada (40.5°N)...	60	80	1962–63	Lake	2.77	0.73	2.55	3.82	902	Yamamoto and Kondo (1964)
				20-cm pan	0.68	1.99	2.67	1.26	604	

[1] Estimated.

Originally, it was believed that the rate of evaporation from a small pan filled with water and placed in the open would be the same as that from a large reservoir. It soon became obvious, however, that this is not the case and that, in fact, evaporation from a pan can far exceed that from a reservoir. Different pans also show different evaporation rates.

Lake and pan evaporation rates are summarized in Table 21 for a few locations in middle latitudes of the northern hemisphere. The values are based on data given by Hughes (1963) for Salton Sea, California; by Harbeck (1958) and Kohler *et al.* (1958)

for Lake Mead, Nevada; by Kohler (1954) for Lake Hefner, Oklahoma; by Dutton and Bryson (1962) for Lake Mendota, Wisconsin; and by Yamamoto and Kondo (1964) for Lake Towada, Japan. Besides lake and standard Class A pan evaporation rates, the table also gives data for a sunken pan, 0.61 m in diameter and 0.92 m deep, covered with 1-cm mesh screen, for Salton Sea and Lake Hefner and for a Class A pan floated on the water surface at Lake Mead. For Lake Towada, values are presented only for the lake and for a small pan, 20 cm in diameter. The data are averaged by seasons: winter (January to March), spring (April to June), summer (July to September), and fall (October to December).

The relationship between lake and pan evaporation is complex. There are basically three variables: the vapor transfer coefficient D_w; the saturation vapor pressure e_s, or the temperature T_s of the water surface; and the vapor pressure e of the overlying air. These are related to the evaporation rate through equation (10.9). Each is discussed below.

If it is assumed that the transfer coefficients for vapor and momentum are equal, it follows from equation (10.20) that

$$D_w = k^2 u\,(\ln z/z_0)^{-2}\,. \tag{11.1}$$

Hence, for a given wind speed, D_w increases with increasing roughness of the surface, which for small bodies of water is determined mainly by the nature of the surrounding terrain. In the case of evaporation pans, the height of the pan rim or water surface above the ground is also an important factor (the rim effect).

Because there is usually a correlation between the wind speed and the vapor pressure difference, $e_s - e$, equation (11.1) cannot be used in equation (10.9) if only average daily values of u, e_s, and e are available. It is easy to show that

$$\bar{E} \propto \bar{u}\,\overline{\Delta e} + \overline{u'\Delta e'}$$

where $\Delta e = e_s - e, u' = u - \bar{u}$, and $\Delta e' = \Delta e - \overline{\Delta e}$. The bar represents a time average of 24 hours. Since the strongest winds normally occur during the day, when the vapor pressure difference is large, and the lightest winds at night, when the vapor pressure difference is small, the correlation term $\overline{u'\Delta e'}$ is usually positive. Therefore, evaporation rates estimated from average daily values of u, e_s, and e using equation (11.1) will be too low. Tanner and Pelton (1960*a*) suggest that rates be computed separately for day and night.

The vapor transfer coefficient for water surfaces can be determined experimentally from equation (10.9) if the evaporation rate and the vapor pressure difference are known. Normally, average daily or monthly values of the evaporation rate and vapor pressure difference are used. The resulting values of D_w are almost linearly related to the wind speed u. That is,

$$D_w = a + bu \,. \tag{11.2}$$

With D_w in cm sec^{-1} and u in m sec^{-1} at a height of 2 m, the coefficient b seems to be about 0.2 for all types of water surfaces, from small pans to large oceans. The coefficient a is extremely variable, however, increasing as the surface roughness increases or as the size of the water body decreases. For most types of terrain, a will be of the order of 0.45 for a standard Class A pan, 0.25 for a sunken pan, and 0.07 for a large lake. Hence, all other factors being constant, evaporation rates at any time of year will be greatest from a Class A pan and least from a large lake.

The saturation vapor pressure at the water surface e_s is a function of its temperature, which in turn depends on many variables, including the evaporation rate. In Figure 42, the annual variations of the air temperature at 2 m, the lake temperature, the lake surface temperature, and the temperature of the water in a Class A pan are shown for Lake Mead, based on data given by Kohler *et al.* (1958). The relative placement of the four curves is fairly representative of conditions at all five lakes listed in Table 21, although the period during which the lake surface temperature T_s is less than the air temperature T varies considerably. For example, the difference $T_s - T$ is negative at Lake Mendota only in April and May and at Lake Towada from April through July.

Through the spring the temperature of shallow water bodies closely follows that of the air. Deep lakes, however, warm slowly, because the heat derived mainly from radiative energy and to a lesser extent from the air must spread through a large volume of water. The surface waters of Lake Mead, for example, warm only 8.5°C between the first week in March and the end of May compared to 13.0°C for the water in a Class A pan. The lake as a whole warms by less than 4°C during this period. Nevertheless, because of the volume of the lake, this heating requires an average of about 250 cal per square centimeter per day, which is a significant portion of what would otherwise be available for evaporation. Hence during the spring this factor tends to suppress evaporation from deep lakes relative to that from shallow pans or ponds.

Conditions remain about the same through most of the summer in humid regions, air and shallow water temperatures being almost equal with lake surface temperatures somewhat lower. In arid regions, however, if the temperature of water in a pan were to continue to increase with that of the air through the summer, evaporation rates would eventually become so large that the energy requirements could no longer be met entirely by radiation. Since the water temperature always changes at a rate com-

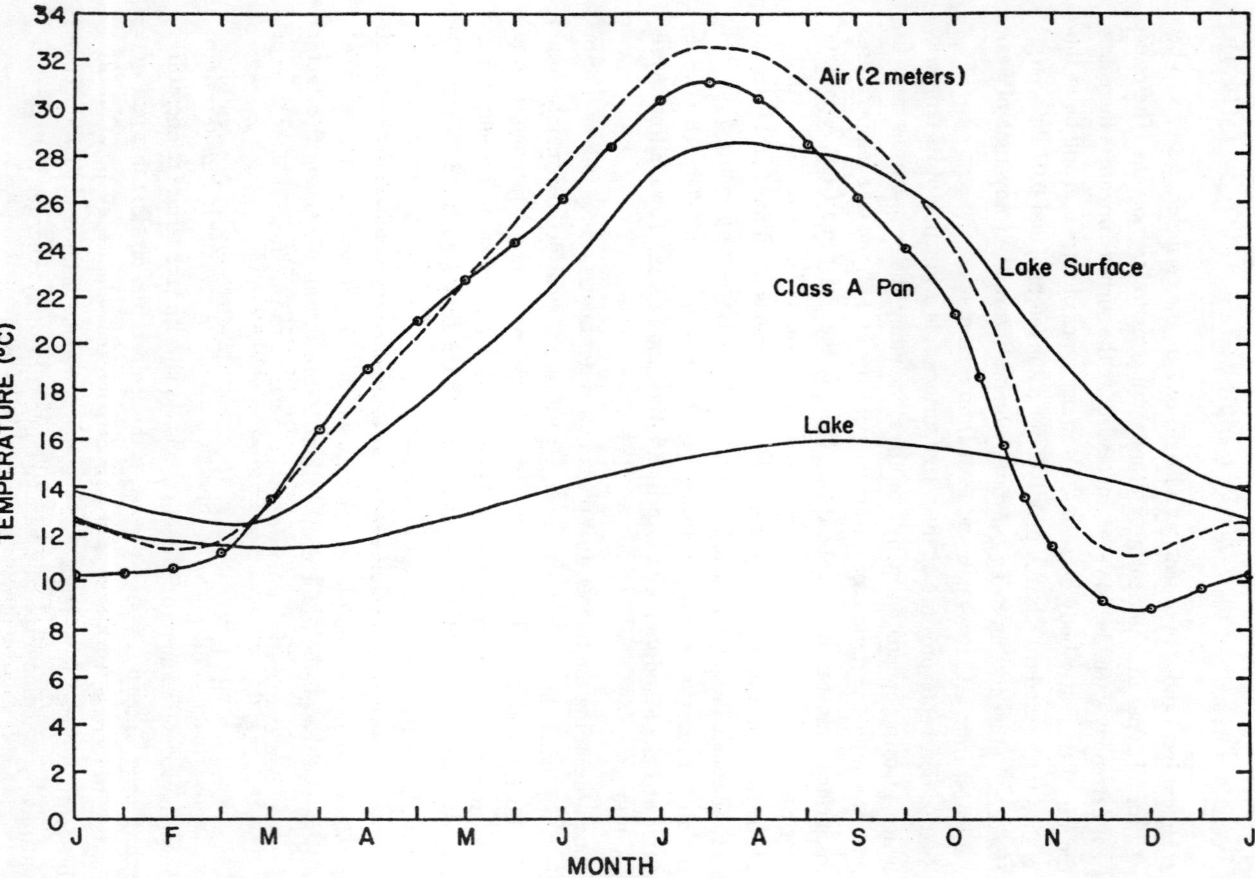

FIG. 42.—The average annual variation of the air temperature at 2 m, the lake temperature, the lake surface temperature, and the temperature of water in a Class A pan for Lake Mead, Nevada. Based on data obtained by Kohler *et al.* (1958) in 1952 and 1953.

patible with the availability of energy for evaporation, during this part of the year pan water temperatures normally average somewhat lower than air temperatures, but still not as low as lake surface temperatures. Furthermore, the existence of a temperature gradient from the air to the surface provides an additional important energy source, the air itself.

In the late summer, when shallow water and air temperatures begin to fall steadily to their winter minima, lake surface temperatures change slowly, as surface waters draw on the tremendous store of heat provided by the lake's interior. Eventually, in the early fall, the lake surface becomes warmer than the air and remains so until the early spring. During this period, then, this factor leads to higher evaporation rates from deep lakes than from shallow water bodies.

This picture is complicated by a number of other factors that can affect water temperatures. In particular, heat transfer through the walls of an elevated pan tends to keep the water warmer than might otherwise be expected when $T_s - T$ is negative. Differences in albedo, associated mainly with differences in pan color or metal type, may also produce significant variations in water temperatures among different types of pans.

The final factor that can affect evaporation rates is the vapor pressure of the air. In humid regions this is likely to be about the same over small and large water surfaces. In arid regions, however, large differences can occur. The desert air is typically dry and has a low vapor pressure. Further, its moisture content is increased imperceptibly as it passes over a pan or other small water surface. In traveling over a large lake, however, air picks up moisture at a sufficiently fast rate to produce a noticeable increase in the vapor pressure by the time the air leaves the lake. At Lake Mead, from October, 1952, to September, 1953, the vapor pressure averaged 10.6 mb near the center of the lake, 8.7 mb on an island near the windward south shore, 7.5 mb at Boulder City, and 4.4 mb at Las Vegas. This factor, frequently referred to as the "oasis effect," should also yield higher evaporation rates from small water bodies than from large ones.

The combination of these three factors, the transfer coefficient D_w, the saturation vapor pressure at the water surface e_s, and the actual vapor pressure of the air e, explains the general features of the data presented in Table 21. In the spring and most of the summer, evaporation from a water-filled pan far exceeds that from a lake; in the fall and winter, amounts are more nearly equal, with a tendency for larger values from the lakes, particularly in the late fall.

The ratio of lake to pan evaporation is frequently called the pan coefficient. It has been used extensively by hydrometeorologists to estimate annual evaporation from proposed reservoirs. A map of the distribution of the average annual Class A pan coefficient over the United States is given by Kohler, Nordenson, and Baker (1959). Values range from less than 0.60 in the Imperial Valley of California to more than 0.80 over the northern Great Lakes and in eastern Maine. Because of the dependence of pan evaporation on pan exposure, maps of this sort should not be used to estimate annual lake evaporation, if better than 10 percent accuracy is required. For obvious reasons, they should never be used to estimate monthly evaporation.

For the lakes listed in Table 21, the pan coefficient ranges from 0.52 for the Salton Sea to almost 1.50 for Lake Towada. The latter is unique in several respects. Not only is its pan coefficient greater than one, but also more than 70 percent of the annual lake evaporation occurs in the fall and winter. In this respect, Lake Towada is similar to the warm ocean currents flowing northward along the eastern coasts of North America and Asia (Fig. 32).

Of the many other types of water evaporimeters we will mention only one, the Piché evaporimeter, which is frequently used in agricultural studies. This evaporimeter is a long, narrow glass tube, graduated in cubic centimeters. One end is closed and the other open and ground flat. The flat end is 1 cm in diameter and covered with a circular filter paper with an area of 13 cm^2. The paper is held in position by a small disk and a metal clip. The tube is filled with distilled water, closed with the filter paper and disk, and inverted. Water soaks the paper and then evaporates from it; the amount of water evaporated in any given time can be measured by making two consecutive readings of the level of water in the tube. The difference gives the volume of water evaporated.

This type of evaporimeter is used for making comparisons over a given area, but does not give absolute results. It is very sensitive to wind speed, and for observations in different places, similar exposures must be used.

Soil evaporimeters.—These are large tanks filled with soil and vegetation, and are used to estimate the maximum amount of water that can be lost from large, well-irrigated fields by evaporation and transpiration. This quantity is called potential evapotranspiration by Thornthwaite (1948) and evaporability by Budyko (1956).

In most models, the tank is irrigated from above with enough water to maintain the soil moisture content above field capacity. The amount of water percolating daily to the bottom of the tank is subtracted from the amount supplied by irrigation or rainfall. The difference is the amount lost by (potential) evapotranspiration. Thus, from the water balance equation,

$$E = \text{water added} - \text{percolation} = r - \Delta f \qquad (11.3)$$

where the storage term is zero because the soil column always has all the water it can hold.

Great care must be exercised in using a soil evaporimeter in arid regions. If it is placed in the middle of a dry field, the evaporimeter will overestimate potential evapotranspiration from a large crop, because the air over the tank is continuously replaced by dry air with a low vapor pressure. By equation (10.9), which applies to any surface, this will tend to keep the vapor pressure gradient at a higher value than it would have with moist surroundings.

Thornthwaite and Mather (1955) suggest that the tanks should have a buffer zone of at least 400 m in deserts to give representative results. In any event, the tanks should be filled with soil and vegetation typical of the surroundings, and the whole area, including the tank, should be kept adequately moist. If the soil in the evaporimeter is allowed to dry, the evapotranspiration rate estimated from equation (11.3) will be less than both the potential rate and the actual rate, since moisture is being withdrawn from the soil. If, on the other hand, the soil outside the tank dries, the measured evapotranspiration rate will be higher than the potential rate because of the oasis effect.

It is also important to keep the vegetation in the evaporimeter level with that in the surrounding field. A difference in the mean height of even 2 or 3 cm could lead to erroneous results. Water loss will be excessive if the vegetation in the tank is higher than that outside. This height difference will influence the transfer coefficient D_w in equation (10.9) which, over land, is very sensitive to changes in the roughness of the underlying surface. The transfer coefficient can increase by about 50 percent with an increase of the roughness length from 0.2 cm (mowed grass, 3 cm high) to 0.7 cm (mowed grass, 7.5 cm high).

An isolated, elevated stand of vegetation (or a tree) presents more surface area to the air than a level stand, increasing both the amount of solar radiation intercepted and the convective heat gain from the air if the vegetation surface is cooler than the air, which will often be the case on warm summer days in arid regions. The ratio of the direct-beam solar radiation Q_E intercepted by an elevated stand to that incident on a horizontal surface Q is given by

$$Q_E/Q = 1 + (h/x) \tan Z \tag{11.4}$$

where h is the height of the vegetation above the surrounding vegetation, x is the width of the vegetation in the direction of the solar azimuth, and Z is the solar zenith angle.

In arid or windy regions the controlling factor in transpiration from vegetation may be the rate at which water diffuses through the stomata, which are usually most abundant on the underside of leaves. Since this rate varies from one plant species to another, this suggests a similar variation in the potential evapotranspiration rate within the same climatic regime.

Lysimeters.—These are large tanks filled with soil and vegetation similar to that of the surroundings and supported on some type of weighing mechanism. These tanks measure actual evapotranspiration and are not necessarily kept at field capacity. Evapotranspiration rates are obtained from weight changes of the lysimeter. According to the water balance equation,

$$E = \text{weight change} + \text{water added} - \text{percolation} = -g + r - \Delta f\,. \quad (11.5)$$

Hence, lysimeters do not measure evapotranspiration rates directly unless the water added and percolation equal zero.

Several factors must be considered in constructing a lysimeter. We must keep heat advection through the walls at a minimum, preferably by using a circular tank and allowing as little space as possible between the lysimeter walls and the surrounding soil. Good thermal contact with the surrounding soil is very desirable, since the vertical temperature gradient should be the same in the two media. Also, to maintain a moisture gradient representative of the surroundings, adequate drainage facilities must be provided at the base of the lysimeter. Otherwise, when precipitation is excessive, water in the tank cannot percolate to the deeper soil layers and accumulates in the lysimeter. Even with proper drainage, under dry conditions the isolation of the lysimeter from the deeper soil layers results in a rapid desiccation of the soil in the tank. One of the biggest functional problems of lysimeters occurs at the boundary of the lysimeter soil and the air beneath it. When the gravitational water reaches this point, it must overcome the resistance set up by surface tension before it can leave the tank.

Typical lysimeters are those at Davis, California, and Tempe, Arizona. The Davis lysimeter, described by Pruitt and Angus (1960), is circular, 6.1 m in diameter, 0.92 m deep, and has a sensitivity of ± 14 kg (± 0.04 mm) with additional errors due to fluctuating wind speeds. The air in a small room under the lysimeter is kept at the same temperature as the surrounding soil at a depth of 1 m in order to produce a realistic temperature gradient. An elaborate drainage system removes excess water. The three Tempe lysimeters, described by Van Bavel and Myers (1962), are square, 1 m on a side and 1.61 m deep. They are sensitive to about 0.01 mm of evaporation or rainfall and, depending on external conditions, they are accurate to 0.02 or 0.03 mm. The ultimate accuracy of lysimeters is limited by the effect that wind has on the weight of an isolated soil column.

If representative evapotranspiration rates are to be obtained, a lysimeter, like an evaporimeter, must be located in the middle of a homogeneous area and must contain soil and vegetation typical of the area.

Two interesting studies of lysimeters are discussed below. In the first, by McIlroy and Angus (1964), evapotranspiration from irrigated grass is compared with that from wet soil and various types of water surfaces. In the second, by Van Bavel, Fritschen, and Reeves (1963), evapotranspiration rates from level and isolated stands of tall Sudan grass are compared.

McIlroy and Angus (1964) used twelve lysimeters, 1.6 m in diameter and 1.1 m deep, to compare evapotranspiration rates from grass, bare soil, and water at Aspendale, Australia, during 1959, 1960, and 1961. Pasture grass was planted in seven of

TABLE 22

SUMMARY OF EVAPORATION DATA OBTAINED BY MCILROY AND ANGUS (1964) AT ASPENDALE, AUSTRALIA DURING 1959, 1960, AND 1961

Method Used to Estimate Evaporation Rate	Evaporation Rate (mm day^{-1})				Annual Total (mm)
	Winter	Spring	Summer	Fall	
Lysimeter					
Grass (6–10 cm)	1.53	5.07	5.88	1.74	1,296
Wet soil	1.05	3.70	4.30	1.07	939
Water	1.34	4.12	5.10	1.46	1,096
Australian tank	1.66	4.37	5.50	2.04	1,242
Class A pan	2.13	5.85	6.77	2.18	1,543
R/L	1.55	4.63	4.64	1.19	1,096

the lysimeters and over an area of about 4,000 m^2 surrounding them. The grass was irrigated as much as four times daily and mowed to a height of 6 to 10 cm. Three of the remaining lysimeters were filled with water and two with bare soil. All were surrounded by moist grass. The results are summarized in Table 22. Values are averaged for winter (July to September), spring (October to December), summer (January to March), and fall (April to June). For comparison, data are also given for an Australian tank, which is a water-filled sunken container about 1 m in diameter and 1 m deep, and for a Class A pan, both located about 20 m from the main irrigated area. The ratio of the net radiation R to the latent heat L is given as an estimate of evaporation in the last row of the table.

The systematically lower evaporation rates from wet soil and water than from grass are attributed mainly to roughness differences. In this case, the partial sheltering of the soil and water lysimeters from the wind by the surrounding grass may have also been an important factor. The higher evaporation rates from the Class A pan than those from either the sunken tank or the grass can undoubtedly be attributed chiefly to the elevation of the pan, whose water surface was about 20 cm above the ground. Since grass evaporation exceeds the ratio R/L in all seasons except winter, part of the energy for evaporation must have been provided by air advected over the field from the drier surroundings.

The study of Van Bavel *et al.* (1963) was conducted on July 23 and 26, 1962. On the first day, green Sudan grass about 100 cm high covered the three Tempe lysimeters and the surrounding field. More than 70 mm of water had been applied to the area on July 19, so that the soil was quite wet and evapotranspiration was probably proceeding at close to the potential rate. During the day, 9.76 mm of water was lost by evaporation. The energy (577 ly) was provided partly by radiation (377 ly, or 65 percent) and partly by convection and advection (200 ly, or 35 percent), since the grass surface was cooler than the air.

On July 26 the grass had been cut to the ground in the surrounding field. This left an isolated stand of Sudan grass, 120 cm high, on one lysimeter. Evapotranspiration reached 14.66 mm, the highest daily rate ever reported in the literature. Since the net radiation on this date was only 399 ly, it is tempting to conclude that most of the energy for evaporation (866 ly) was provided by the air. But, when the radiative energy absorbed through the sides of the stand is taken into account using equation (11.4) and data given by Van Bavel and Fritschen (1964), it appears that, in fact, about 90 percent of the required energy was of radiative origin.

Because of the high evapotranspiration rates, Van Bavel *et al.* (1963) conclude that transpiration from a stand of Sudan grass under conditions of high soil moisture content is fully regulated by external meteorological and morphological factors, physiological factors being unimportant. The grass showed no visible signs of water stress (leaf-edge rolling or change of color), nor was there any depression of transpiration at midday when the evaporative demand was greatest. These authors are careful to restrict their conclusions to Sudan grass and to point out that other crops might behave differently.

CLIMATOLOGICAL METHODS

It is customary to distinguish between two stages of evaporation from soil and vegetation. In the first stage, when the soil moisture content is greater than some critical value w_k, evapotranspiration proceeds at about the potential rate E_o and is dependent mainly on external meteorological factors. In the second stage, when the soil moisture content is less than the critical value, the speed of evapotranspiration depends on the moisture content of the soil, with the relationship usually assumed to be linear. Thus, in the first stage,

$$E = E_o = \text{constant} \qquad w \geq w_k$$

and in the second stage,

$$E = aw \qquad w \leq w_k$$

where $a = E_o/w_k$ and w is the volume fraction of soil moisture, usually expressed in millimeters of water, in the active soil layer. Some authors assume that evaporation from soil covered with vegetation becomes insignificant when the moisture content reaches the wilting point, and represent the second stage by

$$E = a'\,(w - w_{\min}) \qquad w \leq w_k$$

where $a' = E_o/(w_k - w_{\min})$, $w_{\min}$ is the soil moisture content at the wilting point, and $w - w_{\min}$ is the available or productive soil moisture.

There is no general agreement on the value of w_k. Measurements by Alpatev in Russia, quoted by Budyko (1956), indicate that evaporation from field crops is close to the potential value when soil moisture is not lower than 70 to 80 percent of field capacity. On the other hand, Marlatt *et al.* (1961) conclude from field experiments with snap beans planted in a sandy loam soil that evaporation will proceed at the potential rate until about 84 mm of water per meter depth of soil (below field capacity) is lost. This corresponds very closely to a moisture tension of 1 atmosphere and would give values of $w_k/w_{\max}$ ranging from 0.045 for coarse sand to 0.78 for heavy clay. $w_{\max}$ is the soil moisture content at field capacity. Thornthwaite and Mather (1955) assume that potential evapotranspiration occurs only when the soil is at field capacity or above.

More recently, Denmead and Shaw (1962), from field experiments with corn planted in a silty clay loam soil, conclude that the ratio $w_k/w_{\max}$ should increase as the potential evapotranspiration rate increases. The following results can be obtained from their data. The soil has a moisture content of 360 mm m^{-1} at field capacity and

E_o (mm day^{-1})	w_k (mm m^{-1})	$w_k/w_{\max}$
1	232	0.64
2	240	.66
3	253	.70
4	267	.74
5	301	.84
6	335	.93
7	343	0.95

220 mm m^{-1} at the permanent wilting point. For low values of E_o, transpiration rates do not start to decrease until the moisture content is near the permanent wilting point; for high values of E_o, rates start to decrease when the moisture content is almost at field capacity. These results, expressed in terms of the ratio w_k/w_{max}, are not out of line with those of Alpatev.

Denmead and Shaw (1962) further conclude that plant wilting begins as soon as the evaporation rate starts to decrease from its potential value rather than at the permanent wilting point. This conclusion is supported by evidence presented by Kozlowski (1964) from a number of other sources.

There are several bookkeeping methods for estimating evaporation from climatic data. All require some knowledge of the potential evapotranspiration rate.

As pointed out by Thornthwaite (1948), when the root zone of the soil is well supplied with water, the amount used by the vegetation depends more on the amount of solar energy received by the surface and the resultant temperature than on the kind of vegetation growing in the area. (This is not true for some crops; for example, pineapple transpires at night and loses little water during the day.) The water loss under optimum conditions of soil moisture E_o thus appears to be determined principally by climatic conditions and especially by the radiative energy input.

Potential evapotranspiration.—There are many methods for estimating potential evapotranspiration. Only a few of the more widely used or theoretically sound ones are discussed here. The most reliable, and also the most complicated, methods take into account all variables pertinent to the problem: the air temperature, the air humidity, and the available radiative energy. Such an approach has been suggested independently by Penman (1948) and Budyko (1956).

Budyko combines equations (10.8) and (10.9) with the energy balance equation, assuming that $D_w = D_h = D$, to obtain

$$R_o - G_o = A(e_s - e) + B(T_s - T) \tag{11.6}$$

when the radiation balance R_o of the moistened surface is known and

$$R_h - G_o = A(e_s - e) + B'(T_s - T) \tag{11.7}$$

when only the absorbed solar radiation is known and the effective outgoing radiation must be estimated from its relationship to the air humidity and temperature. In these equations

$$A = 0.622\,\frac{\rho L D}{p}, \qquad B = \rho c_p D, \qquad B' = \rho c_p D + 4\epsilon\sigma T^3,$$

and e_s is now the saturation vapor pressure at the temperature T_s of the evaporating surface. The latter two variables, the only unknowns in equations (11.6) and (11.7), are determined by trial-and-error, since they are related through the Clausius-Clapeyron equation. Given one, the other can be obtained from saturation vapor pressure tables, such as those published by List (1958). The sensible heat flux and the evaporation rate then follow from equations (10.8) and (10.9).

Penman eliminates the temperature difference $T_s - T$ from equation (11.6), using the finite difference form of the Clausius-Clapeyron equation,

$$T_s - T = \frac{R_d T^2}{0.622 L e_{sa}} (e_s - e_{sa}) = \frac{1}{\Delta} (e_s - e_{sa})$$

where R_d is the gas constant for dry air and e_{sa} is the saturation vapor pressure at the air temperature. This is a good approximation as long as the temperature difference is small. Then, since $LE_o = A(e_s - e)$, it follows that

$$LE_o = \frac{(R_o - G_o)\Delta + \gamma LE_a}{\Delta + \gamma} \tag{11.8}$$

where $\gamma = c_p p/0.622L$ and $LE_a = A(e_{sa} - e)$.

McIlroy (Slatyer and McIlroy, 1961) modifies the Penman equation using the relationship

$$T - T_w = -\frac{1}{\gamma} (e - e_{wa})$$

and assuming that

$$\frac{e_{sa} - e_{wa}}{T - T_w} = \frac{e_s - e_{sa}}{T_s - T} = \Delta$$

where e_{wa} is the saturation vapor pressure at the wet-bulb temperature T_w of the air. He then obtains

$$LE_o = \frac{(R_o - G_o)\Delta}{\Delta + \gamma} + \rho c_p D_h (T - T_w). \tag{11.9}$$

Over large moist surfaces the sensible heat flux will normally be small (less than 10 percent of LE_o) and, to a first approximation, may be neglected. Then

$$LE_o = R_o - G_o. \tag{11.10}$$

The magnitude of the error due to this approximation is proportional to the temperature difference between the surface and the air. The largest errors will normally occur over small irrigated plots, where the soil surface may be several degrees cooler than the air during the day. In this case, potential evapotranspiration will be underestimated by equation (11.10). This equation, without G_o, has been used earlier in this text.

The air vapor pressures or wet bulb temperatures used in equations (11.6) to (11.9) in arid regions should be those representative of a wet surface and may be considerably larger than the values obtained from published climatological summaries for the area. For example, measurements of dew points over a cotton field and within a citrus grove near Yuma, Arizona, by Ohman and Pratt (1956) indicate that daytime humidities in summer at 1 m above a well-irrigated field may be 40 to 60 percent higher than humidities over the surrounding desert. Hence, use of published mean humidities in the above equations often leads to overestimates of E_o in arid regions, except over small irrigated plots.

The transfer coefficients, D and D_h, can be obtained either from equation (11.1) for calculations based on hourly or half-day values of the pertinent variables, or from equation (11.2) if only daily or monthly average data are available. The coefficients appearing in equation (11.2) are extremely questionable, especially for crops or other tall vegetation, thus limiting the accuracy of the potential evapotranspiration estimates.

Penman (1956), using values of D appropriate for a sunken tank, considers his approach valid only for free water surfaces and suggests that the resulting potential evaporation estimates should be reduced by as much as 40 percent in order to give reasonable estimates of E_o for crops. More recently Penman (1961) acknowledges the importance of surface roughness in increasing evapotranspiration rates. He believes, however, that other factors act to give essentially the same maximum water loss from tall and short vegetation. Tanner and Pelton (1960*a*) disagree, concluding from their studies that "there is ample evidence that evapotranspiration from crops may exceed the evaporation from pans and lakes, particularly when there is appreciable advective heat transfer. Thus, there is no general reason to discard potential evapotranspiration estimates that exceed pan evaporation." The data of McIlroy and Angus (1964) seem to be in accord with this conclusion.

Often it is not practical to use equations (11.6) to (11.10) to estimate potential evapotranspiration, either because the necessary data are not available or because the number of calculations involved is excessive. In such cases, there are a number of simple approximate methods that may be used. The ones to be discussed here are all based on air temperature, since this climatic variable is readily available and, in both moist and arid regions, is essentially independent of the wetness of the surface.

The best-known and most widely used method for computing potential evapotranspiration using air temperature alone was developed by Thornthwaite (1948). Using all available data, he obtained the following empirical relationship between potential evapotranspiration E_o and the mean monthly temperature T in degrees Centigrade.

$$E_o = 16(10T/I)^a \text{ mm mo}^{-1}$$

where

$$a = (0.675I^3 - 77.1I^2 + 17{,}920I + 492{,}390) \times 10^{-6}$$

and

$$I = \sum_{1}^{12} \left(\frac{T}{5}\right)^{1.514}$$

The summation is over the 12 months. These relations are valid only when the air temperature is between 0 and 26.5° C. For lower temperatures, Thornthwaite assumes that the potential evapotranspiration rate is zero. For higher temperatures, the rate is considered to increase monotonically with increasing temperature, being independent of the heat index I. The values of E_o obtained with the Thornthwaite equations must be adjusted to take into account the variation of day length with latitude.

Tables and nomograms for computing potential evapotranspiration by the Thornthwaite method have been published by Thornthwaite and Mather (1957), Van der Bijl (1958), Palmer and Havens (1958), and Van Hylckama (1959). At least one computer program is also available.

In developing and testing his method, Thornthwaite (1948) assumes that water use in selected irrigated valleys in the western United States approximates potential evapotranspiration. This immediately implies that crops are irrigated often enough to guarantee continuous evapotranspiration at the potential rate. Recent unpublished data collected by Van Bavel and Fritschen at Tempe, Arizona, support this conclusion. Their results indicate that, depending on the soil type, evapotranspiration from crops can proceed at the potential rate for more than a month after heavy irrigation, even in midsummer. For bare soil, on the other hand, the rate starts to decrease almost immediately. Most crops, however, are irrigated regularly only during the growing season and must depend on precipitation for their water supply during the rest of the year. Hence, there is at least this one reason for believing that potential evapotranspiration may exceed actual crop water needs at certain times of the year in arid regions.

Penman (1956*a*) touches on the same subject, pointing out that a method suggested by Blaney and Criddle (1962) is concerned, not with potential evapotranspiration, but with crop water needs. On the basis of data gathered from various irrigation projects and lysimeters in the western United States, these authors arrive at a formula that involves multiplying the fraction of the total daylight hours for the year falling in a given month (or on a given day) by the mean air temperature in degrees Fahrenheit. This consumptive-use factor is then multiplied by an empirical consumptive-use crop coefficient to give the water need in inches per month (or per day). The crop coefficient varies both monthly and from one crop to the next, ranging from 0.0 to 1.20. For most crops, average values for the growing season lie between 0.5 and 0.9, with a mean of 0.7. Blaney (1955) obtained a coefficient of 1.15 for evaporation from Silver Lake in the Mojave Desert of California during 1938 and 1939. Although it is crude, the method is widely used in agriculture.

Budyko (1956) has attempted to give physical meaning to the air temperature as a measure of potential evapotranspiration by relating it to the available radiative energy. Using a large number of data, he finds that

$$R_o = 10\Sigma T \text{ ly year}^{-1}$$

where R_o is the annual radiation balance for a wet surface with an assumed albedo of 0.18 and ΣT is the sum of all daily mean temperatures greater than 10°C. In some coastal and mountain regions R_o will be slightly larger than $10\Sigma T$, but normally the approximation is good. Further, the temperature summation is a relatively stable parameter, being about the same over an irrigated field as over a nearby dry desert and, hence, is useful as an index of potential evapotranspiration. Since heat storage may be neglected in the average annual energy balance equation for the earth's surface, then, if it is assumed that the convective heat loss H is also zero, from equation (11.10),

$$E_o \simeq \frac{R_o}{L} \simeq 0.2\Sigma T \text{ mm year}^{-1} .$$

The same equation can be used to estimate potential evapotranspiration for periods shorter than one year. The inaccuracy, however, increases as the length of the period decreases.

Methods for estimating potential evapotranspiration, using the data given by McIlroy and Angus (1964) for Aspendale, Australia, are compared in Table 23. Aspendale is south of Melbourne on the southeast coast of Australia at a latitude of 38°S. Its average annual precipitation is about 630 mm.

In the left-hand portion of the table are given, for each month, the mean air temperature, relative humidity, and wind speed at a height of 1 m, the evaporation rate E_p from a Class A pan, and the evapotranspiration rate E_g from the grass surface. As described earlier, E_g was measured with seven lysimeters containing pasture grass. Note that E_p is greater than E_g in all months.

The size of the grass plot, about 4,000 m², is small for such a study. McIlroy and Angus (1964), however, could find no significant variation of the evapotranspiration rate with distance in from the border of the irrigated area. They attribute the relatively high rates to large-scale advection of dry air from the general surroundings.

In the right-hand portion of Table 23 are given the estimated potential evapotranspiration rates. In the Budyko and Penman methods it is assumed that $D = 0.3(1 + u)$ cm sec^{-1}, where u is in meters per second. In the McIlroy method, $D_h = 0.4(1 + u)$ cm sec^{-1}, which probably accounts for most of the difference between the two sets of estimates. Heat storage in the soil is neglected throughout. In the Blaney-Criddle method, a crop coefficient of 0.8 is assumed.

The Budyko-Penman estimates agree best with the observed evapotranspiration rates, being more than 0.4 mm day^{-1} off only in January and March. They seem to be

TABLE 23

METHODS FOR ESTIMATING POTENTIAL EVAPOTRANSPIRATION FROM SHORT GRASS AT ASPENDALE, AUSTRALIA

(Basic data are from McIlroy and Angus, 1964)

Month	T (°C)	RH (percent)	u (cm sec^{-1})	E_p (mm day^{-1})	E_g (mm day^{-1})	Estimated Potential Evapotranspiration (mm day^{-1})					
						Budyko-Penman	McIlroy	R_o/L	Thornthwaite	Blaney-Criddle	0.2 T
Jan.	23.3	54	258	8.68	7.78	6.86	7.57	6.04	4.47	4.91	4.66
Feb.	21.1	57	245	6.82	5.62	5.44	6.12	4.72	3.51	4.34	4.22
Mar.	19.6	65	227	4.81	4.21	3.76	4.22	3.16	2.80	3.82	3.92
Apr.	17.2	67	223	3.35	2.94	2.76	3.16	2.04	2.04	3.24	3.44
May	12.6	76	223	1.83	1.31	1.48	1.67	0.96	1.09	2.52	2.52
June	10.9	81	218	1.35	0.99	1.01	1.19	0.58	0.80	2.26	2.18
Jul.	10.0	79	230	1.39	0.93	1.17	1.34	0.75	0.74	2.25	2.00
Aug.	11.0	77	212	2.00	1.37	1.60	1.76	1.39	0.89	2.55	2.20
Sep.	13.0	72	238	3.02	2.30	2.64	2.88	2.54	1.50	3.02	2.60
Oct.	15.8	67	268	4.61	4.07	4.00	4.28	3.85	2.09	3.64	3.16
Nov.	17.8	63	265	5.53	5.16	4.76	5.27	4.47	2.73	4.17	3.56
Dec.	20.1	57	257	7.41	5.99	5.99	6.56	5.55	3.47	4.62	4.02
Annual total	......	......	....	1,543	1,296	1,260	1,398	1,095	793	1,257	1,170

slightly low in summer (October to April) and high in winter (May to September). The McIlroy estimates are high in all months except January. The annual total of the McIlroy estimates is about 8 percent greater than that observed. Better agreement could have been obtained with smaller values of D_h in equation (11.9). Obviously, the success of the Budyko, Penman, and McIlroy methods is dependent on the proper selection of the transfer coefficient.

The R_o/L estimates are too low in all months except August and September, indicating that the grass surface was generally cooler than the air and that the evapotranspiration rate was augmented by a downward flux of sensible heat. The Budyko method gives a maximum temperature difference, $T - T_s$, of 1.82°C in January and a minimum difference of 0.22°C in September.

The three sets of estimates based on air temperature alone give poor results. The Thornthwaite values are low in all months, especially in summer. The annual total is only 61 percent of that observed. On the other hand, the Blaney-Criddle and temperature summation estimates are low in summer and high in winter.

Although they may be useful in many regions, the Thornthwaite and other temperature-based methods still give erratic results. [See, for example, the papers of Mather (1954), Drinkwater and Janes (1957), Sellers (1964*b*), and Brutsaert (1965).] More consistent results may be obtained with the following procedure in cases where estimates of potential evapotranspiration must be made for a large number of stations in a climatically homogeneous area (or over a long period of time at a single point). First, calculate monthly values of E_o, using one of the energy balance methods, at a centrally located station where the necessary data are available. The values obtained can then be adjusted to all surrounding stations in the same climatic zone using the simple relationship

$$E_o = aT$$

where T is in degrees Fahrenheit.

The coefficient a varies regionally and seasonally and is a function of the local radiation balance, air humidity, and wind speed. With E_o in millimeters per month and T in degrees Fahrenheit, a ranges from 0.6 in June to 2.9 in January at Aspendale, Australia, from 1.3 in December to 3.1 in May at Yuma, Arizona, and from 0.5 in February to 2.0 in June at Madison, Wisconsin.

For further details on the Thornthwaite and mean temperature methods for estimating potential evapotranspiration, see the excellent paper by Pelton *et al.* (1960).

Actual evapotranspiration.—As stated so aptly by Hastings (1960),

> The possibility of estimating evapotranspiration from climatological data must interest any scientist whose work directly or indirectly involves the soil. Of these individuals the ecologist is apt to believe least that it can be done. His work has acquainted him with so many variables which influence evapotranspiration, he sees daily the effect of so many distinct microenvironments within the soil, that he may display skepticism even toward the notion that general parameters may be used to make remote approximations.
>
> At the other end of the scale the climatologist is apt to view the notion with delight because it affords him the opportunity of working with a relatively simple situation where the number of important variables is few, and they can be reasonably approximated with some degree of regularity.
>
> In between these two extremes lie the soil scientist, the hydrologist, and the geologist named respectively in the order of their decreasing skepticism and the increasing degree to which their disciplines require the simplifying, generalizing, and systematizing of large masses of data.

Only one method for estimating actual evapotranspiration from climatological data is discussed in detail. This is a simplification of a more complex method suggested by Budyko (1963*a*). It is applicable only when the effects of snow accumulation on the ground during the winter can be neglected. According to Thornthwaite and Mather (1957), this will be the case when the average temperature of the coldest month does not fall below −1°C.

In this model the moisture for evapotranspiration is drawn entirely from the root zone of the vegetation or from the upper 10 to 20 cm if the soil is bare. The water balance equation for this active soil layer can be written in the form

$$r = E + \Delta f + g + p = E + S + w_2 - w_1 \qquad (11.11)$$

where the water storage rate g is replaced by the change in moisture content $w_2 - w_1$ of the layer during a selected time interval. The sum of the runoff Δf and the downward percolation p through the base of the layer is replaced by the moisture surplus S. Since the net moisture storage over the period of a year is negligible, the percolated water, which is not available for evapotranspiration, and all of the surplus must eventually run off.

In accord with the discussion at the start of this section, it is assumed that

$$E = E_o \quad \text{when} \quad \bar{w} \geq w_k \qquad (11.12)$$

and

$$E = (\bar{w}/w_k)E_o \quad \text{when} \quad \bar{w} \leq w_k \qquad (11.13)$$

where

$$\bar{w} = 0.5(w_1 + w_2)\,. \qquad (11.14)$$

It is assumed further that the surplus is directly proportional to the precipitation and the soil moisture content. That is,

$$S = br\,\frac{\bar{w}}{w_{max}} \qquad (11.15)$$

where b is an empirically determined constant of proportionality that, according to Budyko (1963), ranges from 0.2 to 0.8 and depends on both the precipitation intensity and the potential evapotranspiration rate. For calculations from average monthly data, the relationship

$$b = \frac{0.8r}{E_o + r} \qquad (11.16)$$

gives good results in both arid and moist sections of the United States. Equations (11.15) and (11.16) are compatible with equation (7.10) if the coefficient a in the latter equation is equal to $0.8\,(E_o + r)^{-1}\,\bar{w}/w_{max}$.

Combining of equations (11.11) to (11.15) gives

$$\bar{w}(1) = \frac{r + 2w_1 - E_o}{2 + b\dfrac{r}{w_{max}}} \tag{11.17}$$

when $\bar{w} > w_k$, and

$$\bar{w}(2) = \frac{r + 2w_1}{2 + b\dfrac{r}{w_{max}} + \dfrac{E_o}{w_k}} \tag{11.18}$$

when $\bar{w} < w_k$.

For this method, w_1 at the start of the first period to be considered must either be known or assumed. Entering this value into equation (11.17), along with r, E_o, w_{max}, and w_k, we get $\bar{w}(1)$ for the first period.

If $\bar{w}(1)$ is greater than w_k, $\bar{w} = \bar{w}(1)$; if $\bar{w}(1)$ is less than w_k, $\bar{w} = \bar{w}(2)$. The value of w_2 for the period is then obtained from equation (11.14) and set equal to w_1 for the following period. The process is repeated to give w_2 at the end of the second period or w_1 at the start of the third period, and so on. If the method is used to obtain E and S from long-term monthly averages of E_o and r, this bookkeeping process is continued until w_2 at the end of each month equals w_1 at the start of the next month.

Should it happen that the computed value of $\bar{w}$ is greater than 0.5 $(w_1 + w_{max})$, it must be assumed that $w_2 = w_{max}$ and $\bar{w} = 0.5\ (w_1 + w_{max})$. On the other hand, if the computed value of $\bar{w}$ is less than 0.5 w_1, let $w_2 = 0$ and $\bar{w} = 0.5\ w_1$. The evapotranspiration rate is then determined from equation (11.12) or (11.13), whichever is applicable, and the surplus from equation (11.11).

To illustrate the application of the method, data of Hoover (1944) and Penman (1956*a*) are used to estimate monthly evapotranspiration and runoff from a dense, deciduous hardwood forest near Coweeta, North Carolina. The results are given in Table 24. It is assumed that the root zone extends to a depth of 150 cm and that the loam and silty clay soil can hold 36 percent by volume of water. Then w_{max} is 540 mm. In accord with the results of Alpatev (Budyko, 1956) and Denmead and Shaw (1962), w_k is taken as 0.75 w_{max} or 405 mm.

The potential evapotranspiration rates in the table are those given by Penman (1956*a*) for Asheville, North Carolina. They have not been multiplied by the empirical crop factor used by Penman and have been adjusted slightly to take into account the temperature difference between Coweeta and Asheville. The precipitation data are averages for the period from November, 1936, to October, 1943. The observed runoff values mentioned later are all adjusted to this 7-year period.

According to Hoover (1944), the soil was at field capacity on April 1 of each year. Hence, to start the application, a value of 540 mm was selected for w_1 in April. This was later changed to 532 mm, in agreement with the value of w_2 obtained in March. Entering 532 mm for w_1 in equation (11.17), we get $\bar{w}(1) = 521$ mm, which is greater than w_k. Hence, $E = E_o = 97$ mm, $\bar{w} = 521$ mm, $w_2 = 510$ mm, and $S = r - E_o + w_1 - w_2 = 63$ mm. The value of w_2 then becomes w_1 for May.

In July, $\bar{w}(1)$ is less than w_k. Hence, $\bar{w} = \bar{w}(2) = 398$ mm, $E = (w/w_k)\,E_o = 147$ mm, $w_2 = 394$ mm, and $S = r - E + w_1 - w_2 = 76$ mm.

TABLE 24

WATER BALANCE DATA FOR THE COWEETA EXPERIMENT STATION, NORTH CAROLINA

Measured Data	Month												Year
	J	F	M	A	M	J	J	A	S	O	N	D	
T (°C)	3.9	5.3	7.5	12.3	16.7	19.9	21.6	21.2	18.2	13.2	7.5	3.6	12.6
E_o (mm)	30	49	70	97	125	147	148	123	84	61	33	25	992
r (mm)	168	176	190	138	112	135	215	152	88	81	112	195	1,762
b	0.68	0.63	0.58	0.47	0.38	0.38	0.47	0.44	0.41	0.46	0.62	0.71	.....
$2+b(r/w_{max})$	2.21	2.21	2.20	2.12	2.08	2.10	2.19	2.12	2.07	2.07	2.13	2.26	.....
E_o/w_k	0.07	0.12	0.17	0.24	0.31	0.36	0.37	0.30	0.21	0.15	0.08	0.06	.....
$2+b(r/w_{max})+(E_o/w_k)$	2.28	2.33	2.37	2.36	2.39	2.46	2.56	2.42	2.28	2.22	2.21	2.32	.....
w_1 (mm)	456	494	516	532	510	458	402	394	382	366	366	398	.....
$\bar{w}(1)$	475	505	524	521	484	430	(398)	.....	.....	.....	.....	427	.....
$\bar{w}(2)$	.....	.....	.....	.....	.....	.....	398	388	374	366	382	(427)	.....
E (mm)	30	49	70	97	125	147	147	116	79	55	31	25	971
S (mm)	100	105	104	63	39	44	76	48	25	26	49	112	791
Estimated runoff (mm)	88	98	100	81	60	53	65	56	40	33	41	76	791
Watershed 17	85	109	105	90	67	41	41	38	25	18	21	44	684
Watershed 18	94	128	128	110	82	53	54	47	32	26	29	54	837

The calculations show that evapotranspiration proceeds at the potential rate from December through June. In July, the soil moisture content falls below the critical value and there is a moisture deficit, since actual evapotranspiration is less than the potential value. The soil is driest in October and early November.

Without knowing how much of the surplus S percolates to the deeper soil layers and how much runs off directly, it is difficult to estimate total runoff for each month. Thornthwaite and Mather (1957), however, have found that for large watersheds only about half of the surplus water that is available in any month actually runs off. The rest is detained on the watershed and is available for runoff during the next month.

Applying this concept to the Coweeta data, we get the estimated runoff values shown in Table 24. These are compared with measured values for two small watersheds with areas of about 0.13 km². The agreement is as good as could reasonably be expected, although the estimated runoff is apparently too low in the winter and spring and too high in the summer and fall. The annual total, 791 mm or 45 percent of the annual precipitation, is in good agreement with that observed.

Hoover (1944) is concerned mainly with the changes in water yield that occur when the vegetation is completely removed from a small watershed. It is expected that runoff will increase, partly because the depth of the soil layer actively participating in the evapotranspiration process is decreased and partly because the potential evapotranspiration rate is reduced. The latter drops by as much as 40 percent because of an increase in the surface albedo and a decrease of the surface roughness, thus decreasing both the absorbed solar radiation and the intensity of turbulence. Small cleared watersheds may also be partially protected from strong winds by the surrounding forested areas.

Hoover found that runoff from the cleared watershed was more than 50 percent greater than from the forested watershed, with the greatest percentage increase in water yield occurring in the fall and the least in the spring. His results can be duplicated with the water balance model if, for the cleared watershed, the depth of the active soil layer is decreased to 15 cm and the potential evapotranspiration rates are reduced to 60 percent of the values for a forested watershed. The results are shown in Figure 43. Note that the great increase in runoff is accompanied by a 40 percent decrease in evapotranspiration.

Uhlig (1954), Thornthwaite and Mather (1957), Albrecht (1962), Turc (Castany, 1963), and Palmer (1965), among others, have proposed water balance models for estimating evapotranspiration and runoff. The Thornthwaite-Mather method will be outlined here because it does not appear explicitly anywhere in the literature and because it has recently been used to estimate the components of the water balance for 13,000 stations covering all of the continental and island areas of the world (Mather, 1964).

The Thornthwaite-Mather model has two parts. First, when precipitation exceeds potential evapotranspiration, it is assumed that

$$E = E_o \quad \text{and} \quad S + w_2 = r - E_o + w_1 \qquad (r \geq E_o).$$

If the sum $r - E_o + w_1$ is greater than w_{max}, it is then assumed that $w_2 = w_{max}$ and that the remainder is surplus. Otherwise, $S = 0$ and $w_2 = r - E_o + w_1$. Second, when precipitation is less than potential evapotranspiration, it is assumed that

$$E = r + (E_o - r)\frac{\bar{w}}{w_{max}} \qquad (r \leq E_o)$$

and that $S = 0$. Then,

$$w_2 = w_1 - (E_o - r)\frac{\bar{w}}{w_{max}} \qquad (r \leq E_o).$$

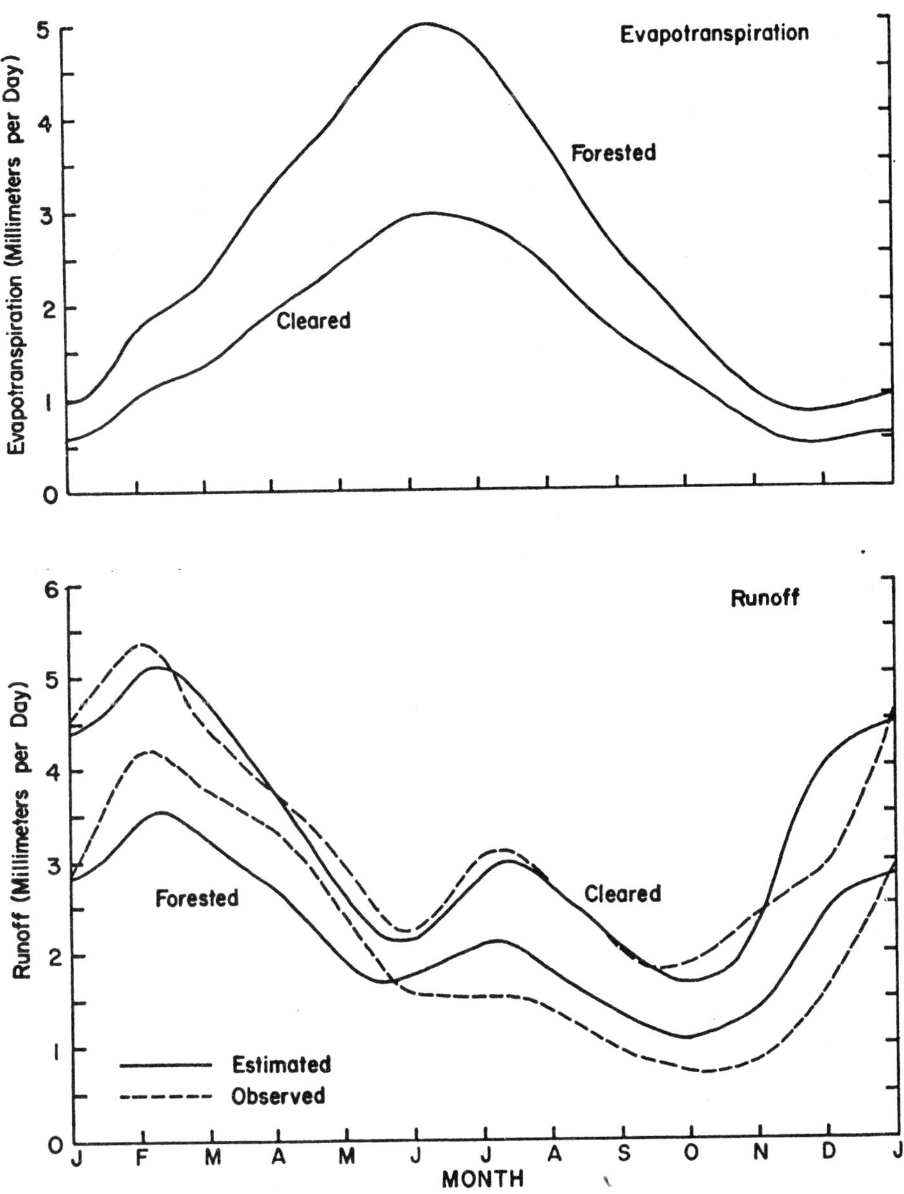

FIG. 43.—Estimated evapotranspiration (*top*) and estimated and observed runoff (*bottom*) from cleared and forested watersheds near Coweeta, North Carolina. The observed data are from Hoover (1944).

During winter months when the mean air temperature is less than $-1°C$, all precipitation is assumed to fall as snow and the water balance equation becomes simply

$$w_2 = r + w_1 \qquad (T \leq -1°C)$$

where there is now no upper limit on either w_1 or w_2. In the first spring month with an average temperature above $-1°C$, all of the excess of w_2 for the preceding month above w_{max} goes immediately into storage and becomes eventual runoff.

This method has two obvious drawbacks. No soil moisture storage is allowed unless precipitation exceeds potential evapotranspiration and no surplus occurs unless the soil is at field capacity. There is also no surplus during winter months when the air temperature is below $-1°C$.

12/ ATMOSPHERIC DIFFUSION

Atmospheric pollution is an important and continuing problem in this age of industrialization, insecticides, and nuclear energy. Anyone who has attempted to spray paint a house on a moderately windy day will readily testify to the ability of particulate matter to travel great distances in the atmosphere and fall where it is least expected and least wanted.

The efficient diffusing capability of the atmosphere is a blessing, as well as an annoyance. For in its absence, the atmosphere surrounding most industrial cities of the world would soon become intolerable for plant and animal life. Under certain atmospheric conditions, gases and small particles accumulate in such quantities that they create a pollution problem and a health hazard. Such events are not uncommon in London, Los Angeles, and parts of the eastern United States. In fact, most cities, at one time or another, have had a smog problem.

The complex theory of atmospheric diffusion has been discussed in detail by Sutton (1953), Wexler (1955), Strom (1962), and Pasquill (1962); therefore, only a few of the more interesting and useful physical results will be discussed here.

The theoretical treatment of atmospheric diffusion usually starts with equation (9.9) written in the expanded form

$$\frac{d\bar{\chi}}{dt}=\frac{\partial}{\partial x}\left(K_x\frac{\partial\bar{\chi}}{\partial x}\right)+\frac{\partial}{\partial y}\left(K_y\frac{\partial\bar{\chi}}{\partial y}\right)+\frac{\partial}{\partial z}\left(K_z\frac{\partial\bar{\chi}}{\partial z}\right) \tag{12.1}$$

where the left-hand member represents the rate of change of the mean concentration of a gas or particulate matter within an identifiable air parcel and K_x, K_y, and K_z are the eddy diffusivities in the directions of the principal axes. The mean concentration $\bar{\chi}$ is usually expressed in grams, curies, or particles per cubic meter.

Consider, first, the relatively simple and idealized case of particles released instantaneously from a point source at the ground into an atmosphere in which neither the eddy diffusivities nor the mean wind speed varies with height. Since both of these variables actually increase rapidly with height close to the ground it should be expected that the results obtained will not be meaningful in the lowest part of the atmosphere. The theory is used most profitably to determine which variables are most important in the diffusion process and to obtain order-of-magnitude estimates of the distance and height of dispersion of airborne substances. The assumption that the eddy diffusivities are constant with height leads to what is called Fickian diffusion.

Assuming that $K_x = K_y = K_z = K$ (isotropic turbulence) and using a coordinate system moving with the speed $\bar{u}$ of the mean wind, we see that equation (12.1) becomes

$$\frac{\partial \bar{\chi}}{\partial t} - c\,\frac{\partial \bar{\chi}}{\partial z} = K\left(\frac{\partial^2 \bar{\chi}}{\partial x^2} + \frac{\partial^2 \bar{\chi}}{\partial y^2} + \frac{\partial^2 \bar{\chi}}{\partial z^2}\right) \tag{12.2}$$

where c is the average terminal velocity of the particles and is always positive.

The appropriate boundary conditions for this problem are that

$$\bar{\chi} \rightarrow 0 \quad \text{as} \quad t \rightarrow 0 \quad \text{for} \quad x, y, z > 0$$

and

$$\bar{\chi} \rightarrow 0 \quad \text{as} \quad t \rightarrow \infty\,.$$

Since matter is neither created nor destroyed during the diffusion process, the continuity equation

$$\iiint_{-\infty}^{\infty} \bar{\chi}\, dx\, dy\, dz = Q \tag{12.3}$$

must hold; Q is the source strength in grams or in number of particles emitted.

Following Roberts (1923) and Rombakis (1947), the solution of equation (12.2) that satisfies the given boundary conditions is

$$\bar{\chi} = \frac{Q}{(4\pi K t)^{3/2}} \exp\left(-\frac{r^2}{4Kt} - \frac{c}{2K}\, z - \frac{c^2}{4K}\, t\right) \tag{12.4}$$

where $r^2 = x'^2 + y^2 + z^2$ and the horizontal coordinates, x' and y, are measured from the center of the particle cloud, which is at a distance $x = \bar{u}t$ downwind from the source at time t. The surface concentration at the center of the cloud is, therefore, given by

$$\bar{\chi} = \frac{Q}{(4\pi K t)^{3/2}} \exp\left(-\frac{c^2}{4K}\, t\right) \tag{12.5}$$

or, since $x = \bar{u}t$, by

$$\bar{\chi} = \frac{Q}{(4\pi K x/\bar{u})^{3/2}} \exp\left(-\frac{c^2 x}{4K\bar{u}}\right). \tag{12.6}$$

A factor of two frequently appears in the numerators of the three preceding equations to take into account the solid earth boundary, which is treated as an impervious reflecting surface. This modification is most applicable to permanently airborne gases or for substances with very low terminal velocities. It is not inappropriate for heavier substances, however, since the main concern is with the matter initially airborne, which theoretically should be only half of that emitted (for a source at the ground). This is equivalent to replacing Q by $2Q$ on the right-hand side of equation (12.3).

Using this concept, Rombakis (1947) shows, by integrating equation (12.4), that the fraction m of the initially airborne particles that are found above the height z at time t is given by

$$m = 1 - \varphi(\zeta)$$

where $\phi(\zeta)$ is the error function,

$$\varphi(\zeta) = \frac{2}{\sqrt{\pi}} \int_0^{\zeta} e^{-v^2} dv \,,$$

and

$$\zeta = (4Kt)^{-1/2} z + (c^2 t/4K)^{1/2} \,. \tag{12.7}$$

Selected values of ζ are listed below for given values of m and $\varphi(\zeta)$. A more complete tabulation is presented by Abramowitz and Stegun (1964) among others.

m	$\varphi(\zeta)$	ζ
0.01	0.99	1.820
.05	.95	1.387
.1	.9	1.163
.2	.8	0.908
.3	.7	0.732
.4	.6	0.595
.5	.5	0.477
.6	.4	0.370
.7	.3	0.273
.8	.2	0.179
.9	.1	0.090
.95	.05	0.044
0.99	0.01	0.008

Solving equation (12.7) for z gives

$$z = \zeta(4Kt)^{1/2} - ct$$

or

$$z = \zeta(4Kx/\bar{u})^{1/2} - cx/\bar{u} \tag{12.8}$$

where z is now the height above which mQ of the particles initially released are found at time t (or at distance x from the source). We can show, following Rombakis (1947), that mQ of the particles are in the air for a period of at least

$$\tau = 4\zeta^2 \frac{K}{c^2}$$

before falling back to the ground. During this time they travel a distance

$$x_{max} = \bar{u}\tau .$$

Enroute, the particles rise to a maximum height of at least

$$z_{max} = \zeta^2 \frac{K}{c}$$

at a distance of 0.25 x_{max}.

In Figure 44, τ and z_{max} are plotted as functions of the terminal velocity c for different values of m and for an eddy diffusivity of 10^4 cm^2 sec^{-1}. The right-hand ordinate gives values of x_{max} for a mean wind speed of 10 m sec^{-1}. Several substances that have terminal velocities between 10^{-2} and 10^3 cm sec^{-1} are listed below the abscissa. These data are taken from Rombakis (1947) and Lapple (1964).

As shown in the figure, particles with low terminal velocities, such as plant spores and insecticide dusts, can remain in the atmosphere for as long as 10 years. Often, however, they are washed out by precipitation long before they would fall naturally to the ground. These small particles are found throughout the troposphere and around the world. Even relatively heavy pollen particles remain in the atmosphere for as long as a day, traveling as far as 1,000 km (600 miles) horizontally and 0.2 km vertically.

For insecticide dusts and gases with terminal velocities of less than 1 cm sec^{-1}, the exponential terms involving c in equations (12.4) to (12.6) usually have an insignificant effect on the observed concentrations at horizontal and vertical distances within 20 km and 20 m, respectively, of the source. Hence, for many small-scale diffusion processes, it is permissible to write equation (12.4) in the form

$$\bar{\chi} = \frac{2Q}{(4\pi Kt)^{3/2}} \exp\left(-\frac{r^2}{4Kt}\right) \tag{12.9}$$

where the factor of two in the numerator takes into account "reflection" from the impervious ground. The surface concentration at the center of a dust or gas cloud is then

$$\bar{\chi} = \frac{2Q}{(4\pi Kt)^{3/2}} . \tag{12.10}$$

These equations were derived on the assumption that neither the wind speed nor the eddy diffusivity changes with height. As mentioned earlier, this is a poor assumption close to the ground and leads to concentration values considerably larger than are actually observed.

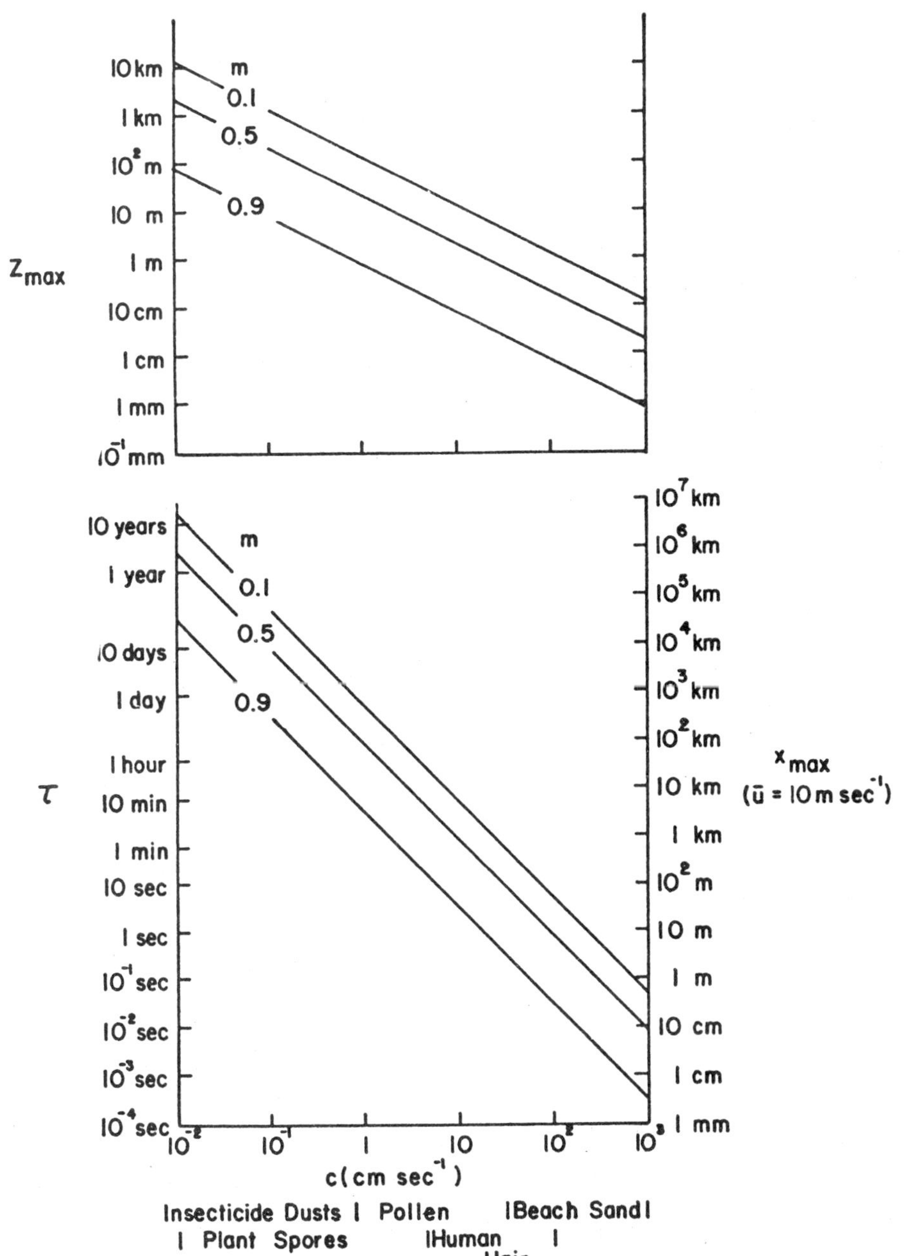

FIG. 44.—z_{max}, τ, and x_{max} as functions of m and c, for an eddy diffusivity of 10^4 cm^2 sec^{-1}

Equation (12.9) is very similar to the trivariate normal distribution for three uncorrelated variables with equal standard deviations σ,

$$\bar{\chi} = \frac{2Q}{(2\pi)^{3/2}\sigma^3} \exp\left(-\frac{r^2}{2\sigma^2}\right). \qquad (12.11)$$

In fact, the two distributions become identical if

$$\sigma^2 = 2Kt = 2Kx/\bar{u}\,.$$

If the particle dispersion is not the same in all three coordinate directions, that is, if the turbulence is not isotropic, it is necessary to distinguish between three different standard deviations, σ_x, σ_y, and σ_z. Equation (12.11) then becomes

$$\bar{\chi} = \frac{2Q}{(2\pi)^{3/2}\sigma_x\sigma_y\sigma_z} \exp\left[-\tfrac{1}{2}\left(\frac{x'^2}{\sigma_x^2}+\frac{y^2}{\sigma_y^2}+\frac{z^2}{\sigma_z^2}\right)\right]. \qquad (12.12)$$

This equation applies to an instantaneous point source release at the ground.

Similar expressions can be written for a continuous point source and a continuous crosswind infinite line source. For a continuous point source

$$\bar{\chi} = \frac{Q}{\pi\bar{u}\sigma_y\sigma_z} \exp\left[-\tfrac{1}{2}\left(\frac{y^2}{\sigma_y^2}+\frac{z^2}{\sigma_z^2}\right)\right] \qquad (12.13)$$

and for a continuous crosswind infinite line source

$$\bar{\chi} = \frac{2Q}{(2\pi)^{1/2}\bar{u}\sigma_z} \exp\left(-\frac{z^2}{2\sigma_z^2}\right). \qquad (12.14)$$

The source strength Q is in units of matter released per unit time in equation (12.13) and in units of matter released per unit length per unit time in equation (12.14). As before, $\bar{u}$ is the mean wind speed in the x-direction. Equations (12.13) and (12.14) correspond, respectively, to the bivariate and univariate normal distributions.

The ground concentration ($z = 0$) at the center of the cloud or along its direction of movement (x', $y = 0$) is for an instantaneous point source

$$\bar{\chi} = \frac{2Q}{(2\pi)^{3/2}\sigma_x\sigma_y\sigma_z}, \qquad (12.15)$$

for a continuous point source

$$\bar{\chi} = \frac{Q}{\pi\bar{u}\sigma_y\sigma_z}, \qquad (12.16)$$

and for a continuous cross-wind infinite line source

$$\bar{\chi} = \frac{2Q}{(2\pi)^{1/2}\bar{u}\sigma_z}. \qquad (12.17)$$

In order to apply these equations, the three standard deviations must be expressed in terms of known or measurable parameters. This has been the topic of recent research in diffusion, particularly by Barad and Haugen (1958, 1959), Barad and Fuquay (1962), Cramer (1959*a*), and Pasquill (1961). The most direct approach is to relate σ_x, σ_y, and σ_z to the standard deviation of the wind direction (azimuth and elevation angle) and to the distance from the source. This has been attempted by Hay and Pasquill (1959), by Cramer (1959*b*), and by Fuquay *et al.* (1964) with some success. Using data from diffusion experiments conducted at O'Neill, Nebraska, in 1956, Cramer presents graphs from which σ_y and σ_z for a continuous point source at the ground can be estimated, given only the standard deviation σ_θ of the horizontal wind direction and the distance downwind. Fuquay *et al.* (1964), on the other hand, working with data collected at Hanford, Washington, conclude that σ_y is more closely related to the product $\sigma_\theta \bar{u}$ than to σ_θ alone. They find, further, that σ_y is proportional to the square of the travel time for times (since release) on the order of a few hundred seconds and proportional to the first power of time for times on the order of thousands of seconds.

Sutton (1953) shows that the relationships

$$2\sigma_x^2 = C_x^2 x^{2-n_x}, \quad 2\sigma_y^2 = C_y^2 x^{2-n_y}, \quad \text{and} \quad 2\sigma_z^2 = C_z^2 x^{2-n_z} \qquad (12.18)$$

are compatible with the statistical theory of turbulence near the earth's surface. C_x, C_y, and C_z are generalized diffusion coefficients, which, along with the three exponents, n_x, n_y, and n_z, must be determined empirically. For Fickian diffusion,

$$n_x = n_y = n_z = 1$$

and

$$C_x^2 = C_y^2 = C_z^2 = 4K/\bar{u}\,.$$

The available measurements of the generalized diffusion coefficients have been summarized by Wexler (1955), Seale and Couchman (1961), and Strom (1962). The data are extremely erratic and confusing, possibly because of the difficulties in directly measuring these parameters. About all that can be said with some confidence is that above 20 to 25 m, turbulence is nearly isotropic and the coefficients are equal. The coefficients vary between 0.05 and 0.30 $m^{n/2}$, decreasing slightly with both increasing height and increasing stability. Below 20 to 25 m, C_x and C_y are apparently greater than C_z and range from about 0.10 to 0.65 $m^{n_y/2}$. Because the vertical component of turbulence is suppressed near the ground, the magnitude of C_z is limited to between 0.04 and 0.25 $m^{n_z/2}$. The change with stability is uncertain. Some of the early studies, summarized by Strom (1962), suggest a variation similar to that observed at higher levels. More recently, Barad and Haugen (1959) and Haugen, Barad, and Antanaitis (1961) found an increase of C_z with increasing stability very close to the ground and little, if any, dependence of C_y on stability. They and Fuquay *et al.* (1963) observe that the magnitudes of the coefficients depend on the roughness of the underlying surface.

The uncertainties surrounding the diffusion coefficients immediately limit the usefulness of the theory. A number of practical results, however, can be obtained without involving the C's at all. Applying equations (12.15) to (12.17) at two points, x_1 and x_2, downwind from the source, using the relationships (12.18) and taking ratios, we get for an instantaneous point source

$$\frac{\bar{\chi}_2}{\bar{\chi}_1} = \left(\frac{x_1}{x_2}\right)^{3(2-n_1)/2}, \qquad n_1 = \tfrac{1}{3}(n_x + n_y + n_z) \tag{12.19}$$

for a continuous point source

$$\frac{\bar{\chi}_2}{\bar{\chi}_1} = \left(\frac{x_1}{x_2}\right)^{2-n_2}, \qquad n_2 = \tfrac{1}{2}(n_y + n_z) \tag{12.20}$$

and for a continuous crosswind infinite line source

$$\frac{\bar{\chi}_2}{\bar{\chi}_1} = \left(\frac{x_1}{x_2}\right)^{(2-n_z)/2} \tag{12.21}$$

Also,

$$\left(\frac{\sigma_{y2}}{\sigma_{y1}}\right)^2 = \left(\frac{x_2}{x_1}\right)^{2-n_y} \tag{12.22}$$

and

$$\left(\frac{\sigma_{z2}}{\sigma_{z1}}\right)^2 = \left(\frac{x_2}{x_1}\right)^{2-n_z}. \tag{12.23}$$

In equation (12.19) the ratio is that of the peak concentration at two points downwind from the source. Equations (12.22) and (12.23) are useful for estimating the lateral and vertical spread of the cloud, since about 68 percent of the total mass should be found within a distance of one standard deviation σ of the cloud center or axis. Almost 95 percent should be found within a distance of two standard deviations.

Sutton (1953) lets $n_x = n_y = n_z = n$ and computes n from the wind profile, assuming that

$$\frac{u_2}{u_1} = \left(\frac{z_2}{z_1}\right)^{n/(2-n)}$$

However, Barad and Haugen (1959) and Haugen *et al.* (1961) present evidence showing that, in fact, n_y and n_z are not equal. They analyze diffusion data collected over prairie grass at O'Neill, Nebraska, during the summer of 1956. Seventy experiments were conducted. In each, sulfur dioxide was emitted continuously for 10 min from a point source 0.5 to 1.5 m above the ground. Gas samplers were mounted at a height of 1.5 m on arcs 50, 100, 200, 400, and 800 m from the source.

The results of twenty-eight of the seventy experiments are summarized in Table 25 and Figure 45. In each release, the gas distribution along each arc was nearly Gaussian or normal. The stability ratio is the temperature difference between 4 m and 0.5 m divided by the square of the wind speed at 2 m. This is a simplified version of the Richardson number, equation (10.24). As shown in Table 25, both n_y and n_z increase with increasing stability, but the former is always considerably larger than the latter. The observed data fit equations (12.20), (12.22), and (12.23) very well, at least for distances between 50 and 800 m from the source. Figure 45 very clearly shows that Fickian diffusion (n_2 = 1.0) does not occur near the earth's surface.

Fuquay *et al.* (1963) summarize the results of a similar set of forty-one half-hour releases in brush-covered desert terrain at Hanford, Washington. In this case, a fluorescent-pigment tracer was used and the sampling arcs were located 0.2, 0.8, 1.6, 3.2, 12.8, and 25.6 km from the source. The data show considerable variability and little dependence on stability. With a stable thermal stratification the axial concentration drops off much more rapidly with distance from the source than observed at O'Neill. This is attributed by the authors to meandering of the plume during the 30-min release period, to the roughness of the terrain at Hanford, and to deposition of the tracer along the plume path.

TABLE 25

ESTIMATED DIFFUSION PARAMETERS FOR O'NEILL, NEBRASKA

(Based on data given by Barad and Haugen [1959] and Seale and Couchman [1961] for 28 gas releases)

Stability ratio (°C sec² cm⁻² ×10⁵)	Number of Cases	n_y	n_z	n_3	σ_y at 100 m (m)	σ_z at 100 m (m)	C_y ($m^{n_y/2}$)	C_z ($m^{n_z/2}$)
3.83.......	4	0.86	0.41	0.64	4.96	2.06	0.45	0.044
0.39.......	11	.64	.20	.43	6.30	3.53	.37	.084
0.11.......	6	.58	.02	.31	7.64	4.32	.38	.068
−0.28.......	7	0.34	−0.19	0.08	11.76	5.16	0.40	0.046

For an instantaneous point source, the total dosage δ received as the cloud passes overhead is of great importance. This can only be determined empirically, since equation (12.12) cannot be integrated directly over time for a fixed value of x. Data collected by Tank (1957) over desert terrain at Dugway, Utah, and presented by Gifford (1957) suggest that the exponent n_3 in

$$\frac{\delta_2}{\delta_1} = \left(\frac{x_1}{x_2}\right)^{n_3}$$

varies from about 0.61 under stable conditions to 1.64 under unstable conditions.

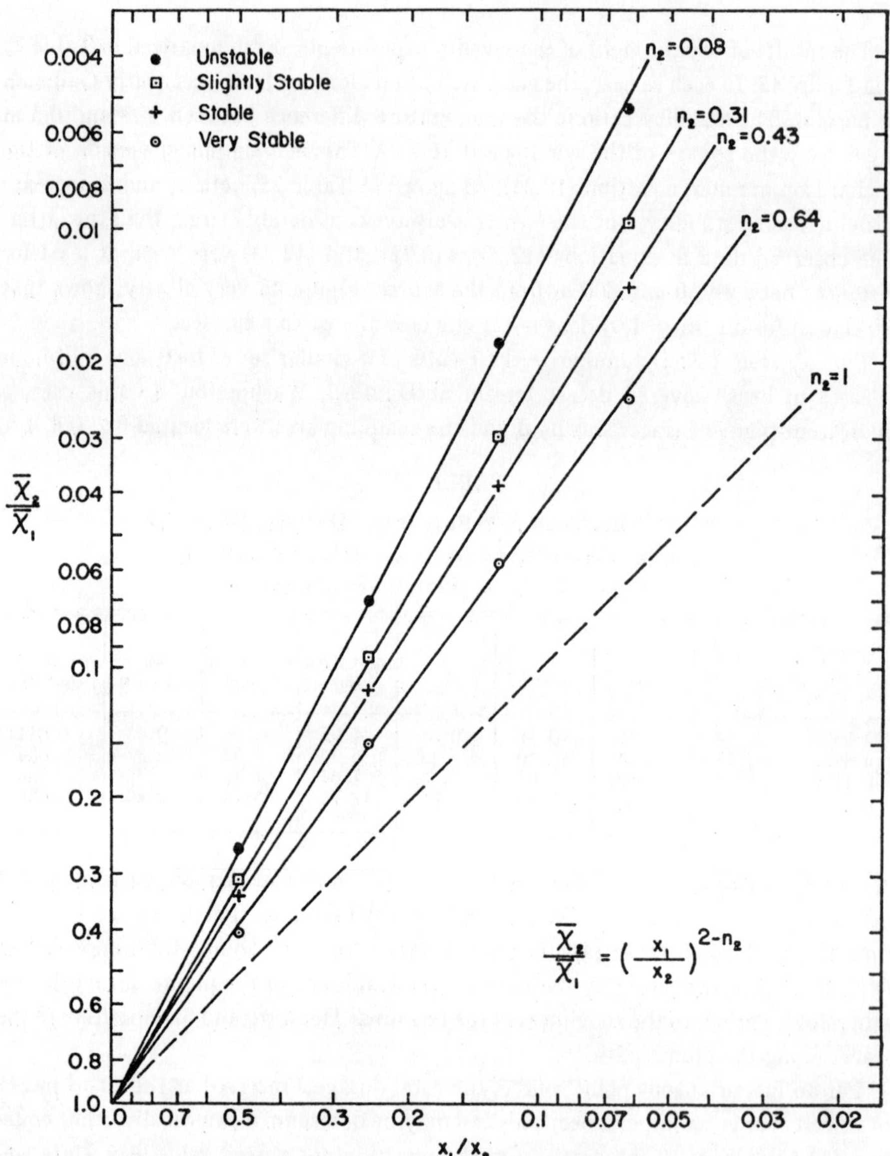

FIG. 45.—The relationship between $\bar{\chi}_2/\bar{\chi}_1$ and x_1/x_2 for a continuous point source at the ground and for different stability conditions. Based on data given by Barad and Haugen (1959) for O'Neill, Nebraska. The dashed line gives the relationship appropriate for Fickian diffusion in which eddy diffusivity does not vary with height.

Probably the most interesting and useful application of diffusion theory is to the continuous emission of gases and particulate matter from elevated point sources. These sources correspond to the tall stacks that are so common in the industrial cities of the world and that discharge several hundred thousand tons of smoke, dust, ash, sulfur oxides, and toxic gases daily into the atmosphere. These contaminants, along with combustion exhausts from motor vehicles, are a byproduct of civilization. They can be controlled to some extent by treatment at the source. The most effective control of stack effluent, however, is obtained by placing the stacks downwind from heavily populated areas and by operating them only when weather conditions are favorable.

The ground-level concentration associated with emission from an elevated continuous point source is given by equation (12.13) written in the form

$$\bar{\chi} = \frac{Q}{\pi \bar{u} \sigma_y \sigma_z} \exp\left[-\frac{1}{2}\left(\frac{y^2}{\sigma_y^2} + \frac{h^2}{\sigma_z^2}\right)\right] \tag{12.24}$$

where h is the height of the plume path or axis above the surface and is usually called the effective stack height. Often it is equated with the true stack height; actually it may be somewhat greater in cases where the effluent is less dense than the atmosphere and possesses an initial vertical velocity. Pasquill (1962) discusses this subject in detail, pointing out that the plume may ascend as much as 100 m and still be rising 1,000 m downwind from the source.

Several interesting geometrical properties of the concentration distribution can be obtained by integrating and differentiating equation (12.24), using the relationships (12.18). The maximum center-line surface concentration is

$$\bar{\chi}_{\max} = \frac{2Q(ah^2e)^{-b}}{\pi \bar{u} C_y C_z} \tag{12.25}$$

and occurs at a distance

$$x_{\max} = (ah^2)^{1/(2-n_z)}$$

downwind from the source. Here

$$a = \frac{2-n_z}{(2-n_2)C_z^2}, \qquad b = \frac{2-n_2}{2-n_z},$$

and e is the base of the natural logarithms (2.718). It is also possible to define a distance x' beyond which the stack height has little effect on the measured concentration. At this point the exponential term in equation (12.24) containing h is nearly one, say 0.95.

Values of $\bar{\chi}_{\max}$, $x_{\max}$, and x' are given in Table 26 for the stability conditions and diffusion parameters listed in Table 25. Data are also given for very unstable conditions with a stability ratio of -1.58. In the latter case it is assumed that n_z, n_2, C_y, and C_z equal −2.08, −0.79, 1.79 $m^{n_y/2}$, and 0.0009 $m^{n_z/2}$, respectively. These values are in accord with data presented by Cramer (1959*b*), Haugen *et al.* (1961), and Seale and Couchman (1961) for O'Neill, Nebraska.

As indicated in Table 26, the maximum center-line concentration depends mainly on the stack height. Since the exponent b in equation (12.25) averages close to 0.85, except under very unstable conditions, it follows that

$$\bar{\chi}_{max} \propto h^{-1.7} .$$

The dependence of $\bar{\chi}_{max}$ on stability is weak and erratic and would be completely absent if both b and the ratio C_z/C_y equal one.

TABLE 26

ATMOSPHERIC STABILITY AND STACK HEIGHT EFFECTS ON DIFFUSION PARAMETERS

Stability Ratio (°C sec³ cm⁻² × 10⁵)	Stack Height (m)								
	$\bar{\chi}_{max}$ (g m⁻³ × 10⁴)			x_{max} (km)			x' (km)		
	30	50	100	30	50	100	30	50	100
3.83	34	14	4.3	4.05	7.70	18.4	23.7	45.2	108
0.39	54	22	6.6	0.74	1.31	2.8	3.6	6.3	13.8
0.11	56	23	7.2	0.51	0.85	1.7	2.1	3.5	7.1
−0.28	30	12	3.6	0.39	0.63	1.2	1.4	2.3	4.3
−1.58	21	10	4.0	0.18	0.23	0.32	0.33	0.43	0.60

$Q = 100$ g sec⁻¹.
$\bar{u} = 5$ m sec⁻¹.

On the other hand, the distance downwind to the maximum concentration is very much dependent on stability. With a strong inversion, most of the effluent remains at its original level and surface concentrations are negligible within fifty stack heights of the source. Lowry (1951) mentions a case where a smoke trail emitted from a 116-m stack at Brookhaven National Laboratory, New York, was followed for 22 miles without any measurable change in elevation or vertical thickness. Under very unstable conditions, the maximum concentration is often found within three stack heights of the source. Fortunately, the mixing and turbulence are so great that maximum average concentrations are usually not excessive.

The height of the stack has its greatest effect on x_{max} under very stable conditions. Changing the height from 30 to 100 m increases x_{max} by more than 14 km (almost 9 miles). Under very unstable conditions, the same height change increases x_{max} by only 150 m (0.1 mile).

The distance x' is even more dependent on stability than is x_{max}. Under stable conditions, the surface concentration is affected by the stack height to distances of more than 1,000 h from the source. On the other hand, under very unstable conditions the stack height has little influence on the surface concentration at distances of more than 10 h. Under these conditions the source might as well be on the ground.

Because of the statistical nature of the basic diffusion equations, the results apply only to long-term mean concentrations and to conditions where the average wind direction does not change appreciably with either time, height, or horizontal distance. The theory can be expected to give the best results when compared with average concentrations obtained from a large number of diffusion experiments carried out under essentially identical weather conditions. Individual releases may not even closely approximate the theoretical expectation.

In the qualitative discussion of plume behavior and its dependence on atmospheric stability, three basic patterns and three transitional patterns are frequently mentioned. These are illustrated in Figure 46. The basic patterns are looping, coning, and fanning, which occur, respectively, when the temperature decreases rapidly with height (unstable lapse conditions), decreases only slowly with height or is constant (near-neutral conditions), and increases with height (stable inversion conditions). The transitional patterns are lofting, fumigation, and trapping. Lofting occurs when the stack extends through a surface inversion to an unstable layer above. Fumigation and trapping occur when an unstable or near-neutral surface layer is capped above the stack by an inversion.

Looping occurs only during daylight hours, usually when the sky is clear and there is little wind and a pronounced super-adiabatic lapse rate. Because of the great instability and convective mixing, the effluent diffuses rapidly and erratically with unusually large concentrations frequently observed momentarily near the base of the stack. Accordingly to Bierly and Hewson (1962), for a 100-m stack the 1-hour average peak concentration associated with looping can be estimated from

$$\bar{\chi}_L(\text{g m}^{-3}) = 4.0\, Q\bar{u}^{-1}\,(x^2 + h^2)^{-0.9}\,. \tag{12.26}$$

This equation gives center-line concentrations at least five times greater than those obtained with equation (12.24).

Coning is most common when it is cloudy and windy and the temperature is nearly constant with height. It occurs more frequently in moist climates than in dry climates and is associated primarily with mechanically induced turbulence. The conical shape of the plume is very nearly that prescribed by the theory, and measured concentrations are usually in accord with those obtained from the standard diffusion equations.

Fanning is observed most often at night and in the early morning when skies are

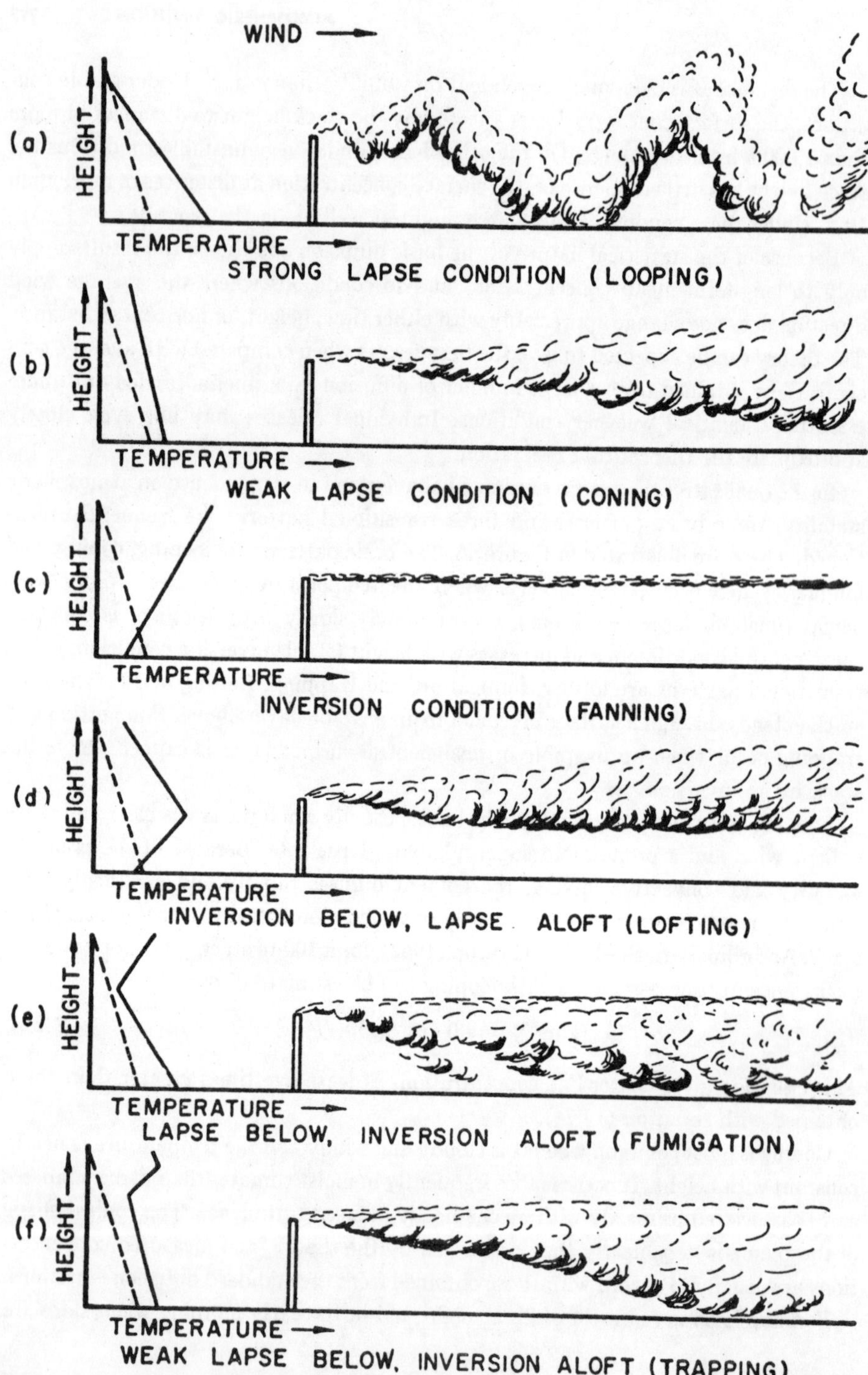

FIG. 46.—Six types of plume behavior under various conditions of stability and instability. The broken lines at left are dry adiabatic lapse rates. The solid lines are existing lapse rates. From Bierly and Hewson (1962).

clear and winds are light, encouraging the formation of a deep, stable air layer near the ground. These conditions may persist throughout the day in higher latitudes, especially if there is a snow cover. The plume, when seen from above, often resembles a meandering river. Turbulence is suppressed and the flow is quasi-laminar. In open country the plume can be tracked for tens of kilometers. Fanning is usually considered a favorable condition for stack releases, even though the effluent is only slowly diluted as it moves away from the stack. A serious pollution problem might occur, however, if the stack is short relative to surrounding objects or terrain features. The concentration at the center of a fanning plume 1 to 2 km downwind from the source is of the same magnitude as the ground-level concentration measured 100 to 200 m downwind from a 100-m stack under looping conditions.

In the early evening on clear, nearly calm, nights it is customary for a surface inversion to form. The inversion starts as a very shallow stable layer near the ground and increases in depth as the night progresses. For a period of 1 to 3 hours near sunset the top of the inversion may be below stack height and cause the emitted plume to spread much more rapidly upward than downward. Except when the inversion is very shallow, this lofting condition may be regarded as the most favorable diffusion situation. The effluent is rapidly diluted in the lapse layer above the inversion, and essentially none of it reaches the ground. The greatest concentrations often occur at the top of the inversion, rather than along the plume axis.

Bierly and Hewson (1962) distinguish between three types of fumigation. The first, and most common, type is associated with the burning off of a surface inversion shortly after sunrise. Solar heating of the ground causes a shallow, unstable layer that grows in depth as the day progresses. When the top of the layer reaches stack height, the effluent is brought to the ground suddenly in high concentrations. Fortunately this condition usually persists for less than an hour. The second type of fumigation occurs on clear, nearly calm, evenings when radiationally cooled and stable air from a rural area moves over an urban area, which acts as an artificial heat source long into the night. The third type of fumigation is associated, for example, with air flow from cool water to warm land during the day in the spring and summer and from cool land to warm water during the night in the autumn and winter. Here the heat source is a natural one.

According to Holland (1953) and Bierly and Hewson (1962), fumigation concentrations at distances of more than 100 m from a 100-m stack can be estimated from

$$\bar{\chi}_F(\text{g m}^{-3}) = 0.28\, Q\bar{u}^{-1}\, x^{-0.75}\,. \tag{12.27}$$

It then follows from this equation and equation (12.26) that fumigation concentrations are seventeen times greater than looping concentrations at 100 m from the source and almost 250 times greater at 2 km. Obviously, when such conditions exist, all stacks should be shut down.

The last diffusion pattern, plume trapping, occurs when an upper-level inversion physically traps the effluent from a stack in the surface air layers. In a sense, this is similar to fumigation. However, fumigation is associated with an inversion of radiative origin which normally disappears by mid-morning. Trapping, on the other hand, is associated mainly with subsidence inversions, which may persist for months, and to a lesser extent with frontal inversions, which usually last less than a day. Subsidence inversions occur most often on the eastern edges of semipermanent high-pressure areas. Here the air aloft is moving equatorward, descending, and warming adiabatically. The situation is aggravated if the air near the surface is cooled from below by contact with either relatively cold ocean water or a snow cover.

In the United States, subsidence inversions and plume trapping occur frequently along the Pacific Coast. Pollution problems are most evident in the Los Angeles basin. Here the vertical dispersion of contaminants is limited by the inversion layer and the lateral dispersion by high mountains to the north and east. The worst conditions occur when the man-made pollution accumulates over a period of several days in calm weather and the base of the inversion layer descends to its lowest level and brings the polluted air to the ground.

Because it can be so persistent, plume trapping is the most dangerous and deadly of all the diffusion patterns and was responsible for the disastrous episodes which occurred in the Meuse Valley in Belgium in 1930; in Donora, Pennsylvania, in 1948; and in London, England, in 1952 (Goldsmith, 1962). Each location is within a heavily industrialized valley and in each case light winds, fog, and a temperature inversion persisted for at least 5 days. This situation permitted contaminants to accumulate to unprecedented levels. Smoke, sulfur dioxide, and possibly fluorides were the main pollutants.

13/ PALEOCLIMATOLOGY AND THEORIES OF CLIMATIC CHANGE

Paleoclimatology is the study of past climates. McDonald (1955) has stated that its ultimate goal is "the complete determination of the variations of climate for all parts of the world and for all portions of the history of our planet, beginning with the time of its formation." Since even present climatic variations are not completely understood, it very doubtful whether this goal will ever be reached. Its attainment is certainly not in sight today.

Nevertheless, if all the resources and reliable results of the various branches of science interested in climatic change are pooled, there comes to light a somewhat hazy, but intelligible, picture of the past. This picture is primarily one of warmth, at least over the 18 percent of the earth lying north of 40°N. According to Brooks (1951), average temperatures for this region were approximately 10° higher than today during more than 90 percent of the past 500 million years.

Periodically, at intervals of about 250 million years, huge ice sheets have spread out from the polar regions, covering as much as 40 percent of the earth's land surface. These glacial epochs, which are accompanied by temperature drops approaching 25° in middle latitudes, resemble nothing so much as periodic and short-lived—to the extent that a few million years can be called short-lived—breakdowns in the earth's heating system. The latest glacial epoch, called the Pleistocene, started about one million years ago and continues today near the poles. Most of man's stay on earth has occurred during this anomalously cool period.

The ice sheets of a glacial epoch are far from stationary. At times they may advance within 40° of the equator over the continental interiors; at other times they may exist only in isolated patches near the poles. The advances have been termed ice ages, the retreats interglacial ages or simply interglacials. Within the Pleistocene epoch there have been at least four ice ages, the latest of which ended about 11,000 years ago in the United States and 9,000 years ago in Sweden.

As our knowledge of the climatic sequence during the past has increased, more and more emphasis has been placed on finding a rational explanation for the observations. As a result, the literature has been saturated with theories of climatic change. It is the purpose of this chapter to review several theories that have been widely, but not universally, accepted. All are valid to the extent that they give results which are not completely contrary to observations or to basic physical principles. Hence, they are all plausible explanations of at least a portion of the spectrum of climatic change. Perhaps it will eventually be proved that one of these processes, or some combination of them, is the principal regulator of climate. But at this stage, uncertainty is the only thing that is certain.

A brief summary of the climatic sequence of the geologic past is presented in the first section of this chapter. No claim is made for absolute accuracy in most of the dates quoted. On the contrary, it should be emphasized that the chronologies presented are, at best, first approximations.

CLIMATES OF THE PAST

For reference purposes, the dates of the various geologic eras, periods, and epochs are given in Table 27. The dates are only approximate, and represent slight modifications of those obtained by Kulp (1961) from isotopic age measurements on rocks of known stratigraphic age. Also shown in the table are the time and extent of known major glaciations. This information is based on data summarized by Schwarzbach (1963).

The Pre-Archeozoic era dates from 4,600 million years B.P. (before the present), the estimated age of the solid earth mantle (Kuiper, 1957), to 2,800 million years B.P., the estimated age of the oldest known rocks (Hurley, 1959). The age of the earth itself, according to Kuiper (1957), is between 5,000 and 6,000 million years.

Very little is known about the climate before the Cambrian period. Wolbach (1955) states that there were at least five pre-Cambrian glaciations, occurring at intervals of 250 million years. The most extensive of these probably took place in the Huronian period, when glaciation was extensive in North America and South Africa. There is evidence, however, that by the late or upper Huronian, a warm tropical climate existed in Canada and the north-central United States. This is not inconsistent, since the so-called glacial epochs rarely lasted longer than a few million years. The total length of the Huronian period was probably about 250 million years.

The Cambrian period, centered at about 550 million years B.P., started with a glacial epoch and ended with a climate warmer than what prevails today. According to Brooks (1951), glaciation was unusually extensive, with glaciers occurring in the temperate zones of North America, tropical South Africa, India, China, and eastern and southern Australia. By the time of the upper Cambrian, a warm climate is indicated for most of the northern hemisphere. Tropical conditions prevailed in California, and dry desert conditions in most of the Midwest (Stoval and Brown, 1955). By the end of the Cambrian period, average temperatures north of 40°N probably were between 10 and 20°C, values typical of tropical regions today.

The warmth of the upper Cambrian continued into the Ordovician period. The climate of most of North America was milder than the present. There are indications that parts of the eastern United States were very dry during the middle of the Ordovician period.

TABLE 27

GEOLOGIC ERAS, PERIODS, EPOCHS, AND REGIONS OF KNOWN MAJOR GLACIATIONS

Era	Period	Epoch	Beginning of Interval (Millions of Years Ago)	Major Galciations							
				Antarctica	South America	Africa	Australia	Asia	Europe	North America	Arctic
Pre-Archeozoic	...	...	4,600	...	...	...	...	...	...	...	...
Archeozoic	Keewatin	...	2,800	...	...	...	...	...	...	x	...
	Timiskaming	...	...	...	...	...	...	...	...	x	...
Proterozoic	Huronian	...	1,000	...	?	x	x	...	...	x	...
	Algonkian	...	...	...	...	...	...	...	...	...	...
	Pre-Cambrian	...	...	...	...	x	x	x	x	?	x
Paleozoic	Cambrian	...	600	...	...	...	...	...	...	...	...
	Ordovician	...	500	...	...	...	...	...	...	...	...
	Silurian	...	430	...	...	...	...	...	...	...	...
	Devonian	...	400	...	x	x	...	...	...	...	...
	Carboniferous	Mississippian	350	...	...	...	...	...	...	...	...
		Pennsylvanian	330	?	x	x	x	x	...	...	...
	Permian	...	275						...	...	...
Mesozoic	Triassic	...	225	...	...	...	...	...	...	...	...
	Jurassic	...	180	...	...	...	...	...	...	...	...
	Cretaceous	...	135	...	...	...	...	...	...	...	...
Cenozoic	Tertiary	Paleocene	66	...	...	...	...	...	...	...	...
		Eocene	59	...	...	...	...	...	...	...	...
		Oligocene	38	...	...	...	...	...	...	...	...
		Miocene	25	...	...	...	...	...	...	...	...
		Pliocene	12	...	...	...	...	...	...	...	x
	Quaternary	Pleistocene	0.6	x	x	...	x	x	x	x	x
		Holocene	0.01	...	...	...	...	...	...	...	...

The Silurian period was also warm, particularly in all of North America and in the Arctic regions (Dunbar, 1960). Geologic evidence suggests that most of the seas that had covered North America in the Cambrian and Ordovician periods disappeared in the upper Silurian (Cayugan epoch), leaving the continent very arid, especially in the eastern United States. The orogenesis of the Caledonian system of northwest Europe, which occurred from 370 to 450 million years B.P., was most active toward the end of this period (Schwarzbach, 1963).

The Devonian period was generally warm, except for local glaciation in South America and South Africa. Indications of glaciation at this time in the northern hemisphere and in Austrialia are considered very doubtful by Schwarzbach (1963). The eastern United States probably had a warm, uniform climate with moderate seasonal rainfall.

The Carboniferous period, which started warm and moist and ended with nearly glacial conditions, is usually divided into the Mississippian and Pennsylvanian periods, here referred to as epochs. The former was warm over most of North America, including Alaska, with summer rains indicated in the eastern United States. Local aridity probably prevailed on the leeward sides of the mountain formations of the western United States. The general trend, however, was toward unusually heavy precipitation, especially in the eastern United States, Europe, and northern Asia. Most of the coal deposits in these areas were formed at this time. These conditions continued into the Pennsylvanian epoch, but by the end of this period the climate was drier and colder than it had been at any time since the Pre-Cambrian glaciation. In Europe, arid continental conditions were enhanced by the Variscan orogenesis from 210 to 310 million years B.P.

During the first quarter of the Permian period, glaciation occurred in southern South America, South Africa, India, and most of Australia. Many authors believe that these regions were physically connected at this time, forming one great land mass called Gondwanaland. According to Wolbach (1955), there may have been from three to five major stages of glaciation, somewhat similar to those that took place during the Pleistocene epoch. The period as a whole can be characterized as generally arid and cold in the southern hemisphere and dry and warm in the northern hemisphere. It was a time of ocean regression and enlargement of land areas. Strangely enough, no glaciation was evident on the coast of Antarctica during the Permian period (Wolbach, 1955). Most of the northern hemisphere continents were also ice-free. Arnold (1947) reports a semiarid climate in the central United States, with torrential rains being followed by intervals of extensive drought. Coal deposits suggest a warm tropical climate in northern China and parts of eastern Australia during much of the period.

The first period of the Mesozoic era, the Triassic, started mild and arid and ended warm and moist, particularly in Tasmania, Australia, Siberia, southern Asia, Europe, eastern Greenland, and Spitzenbergen. A warm climate is also indicated in Alaska, California, and Malaysia by the presence of corals in the adjoining waters.

The Jurassic period, sometimes called the "Age of Corals," was also warm, with temperatures averaging close to 10°C north of 40°N. Sea water temperatures poleward of 50°N were about 5 to 10° higher than they are today. The melting of the Permian glaciers caused much of Europe, Asia, and the western United States to be overrun by sea water. A warm tropical climate prevailed in Japan, Manchuria, Spitzenbergen, arctic Alaska, England, Siberia, and the region south of Cape Horn (Arnold, 1947). Precipitation was, for the most part, intermittent and light.

Oceanic transgression and a uniformly warm climate continued through the Cretaceous period (Dunbar, 1960); however, there was a definite trend toward worldwide cooling following a temperature maximum of 20°C in Europe and North America about 85 million years ago (Emiliani, 1958). Some authorities believe that this cooling led to the extinction of dinosaurs. Coal deposits indicate a warm and wet climate in Alberta during the early Cretaceous and in the western United States during the late Cretaceous. According to Urey *et al.* (1951), mean water temperatures were close to 18°C between 52 and 56°N along the coasts of England and Denmark. Current mean temperatures average 10°C.

The Alps, Caucasus, Himalayas, and the North American Cordillera were formed during the Tertiary period (Öpik, 1957), which is usually divided into five epochs. The earliest of these, the Paleocene, started about 66 million years ago and had a moderate to warm climate and somewhat more precipitation than today. Brooks (1951) reports local glaciation in the Antarctic and the North American Cordillera at the start of this epoch.

Warm and moist conditions continued through the Eocene epoch. Both the average annual temperature in central Europe and the water surface temperature of the northern California coast in February were about 15° higher than they are today (Flint, 1957). The hot summers and mild winters, particularly in high latitudes, precluded ice formation in any part of the world (Brooks, 1951). Most of Europe enjoyed a typically Mediterranean climate, with the heaviest rains falling in winter. In England the climate was similar to that of a tropical rain forest. All of the United States was warm and wet, the western half of the country receiving annually about 180 cm of precipitation evenly distributed through the year (Brooks, 1951). This climate produced luxuriant vegetation and abundant wildlife in regions that today are barren and unproductive. For example, palms, laurels, and magnolias were found in southeastern Alaska; spruce, pine, and willow in Grinnell Land, within 10° of the North Pole; and swamp forests in Tennessee (Dorf, 1960). Alligators ranged as far north as Wyoming and the Dakotas.

The Oligocene epoch also started warm and moist, but by its culmination the climate turned cold and arid, resembling what prevails today. Emiliani (1957) quotes 10°C as the average annual temperature around the Antarctic continent during the early part of the epoch. The tropical vegetation of the Eocene was gradually replaced by temperate forests in most of the northern United States (Dorf, 1960).

A slight amelioration of the climate occurred in Miocene times, representing a temporary reversal of the general trend toward glaciation. Air temperatures in central Europe and water surface temperatures along the northern California coast in February were, respectively, about 10° and 7° higher than they are today. Water temperatures, however, were 3° to 4° lower than they are at present in the equatorial Atlantic Ocean (Emiliani, 1957).

The Pliocene epoch was marked by a sharp downward trend in both temperature and precipitation. Desert vegetation developed in the Great Basin and the interior of northern Mexico because of lower rainfall and greater seasonal temperature contrasts (Dorf, 1960). By the end of the epoch, the climate was both colder and drier than it is today. This led to the onset of the most recent and best known of the glacial epochs.

Although the Pleistocene epoch lasted only one million years, an almost insignificant length of time, it has received much more attention than all of the 4,599 million years of geological time that preceded it. This is obviously because of the proximity of the Pleistocene to the present. Not only are relics of this epoch more readily available, but also a wider variety of dating techniques may be used for this period. In spite of these factors, or perhaps because of them, no consistent geological calendar for the past million years, or even for the past 50,000 years, has been constructed. Published results are often contradictory and confusing. Much of the confusion has apparently resulted from applying cross-dating techniques to regions that, contrary to a widely used assumption, have not undergone the same sequence of climatic events. The fact that climatic anomalies in Europe and the United States have generally been of the same sign during the past 300 years does not make valid the assumption that they have also been of the same sign for the past 300,000 years. The situation is complicated further by the lack of a completely reliable technique for dating events that occurred between 150,000 and 10,000,000 years ago.

A tentative chronology of the ages and subages of the Pleistocene epoch from its start until about 10,000 years B.P. is presented in Table 28. For simplicity, it has been assumed that the glacial ages of higher latitudes were accompanied by pluvial ages in low latitudes and that the ice ages occurred simultaneously in Europe and North America. Because of the wide range of dates given in the literature for the peaks of the various ages and subages of the Pleistocene, the mean date is followed by an estimate of the standard error of this mean. Thus, the Nebraskan glacial, with about 70 percent probability, had its peak sometime between 275,000 and 925,000 years ago. The large values of the standard errors indicate the questionable accuracy of the geological chronologies presented thus far.

TABLE 28

AGES AND SUBAGES OF THE PLEISTOCENE EPOCH

Age	Subage	Thousands of Years Ago
Nebraskan (Gunz) Glacial		
Kageran Pluvial		600 ±325
Aftonian (Gunz-Mindel) Interglacial		500 ±270
Kansan (Mindel) Glacial		
Kamasian Pluvial		420 ±240
Yarmouth (Mindel-Riss) Interglacial		300 ±130
Illinoian (Riss) Glacial		
Kanjeran Pluvial		187 ± 76
Sangamon (Riss-Würm) Interglacial		106 ± 36
Wisconsin (Würm) Glacial		
Gamblian Pluvial	Iowan, Farmdale (Pre-Brandenburg) Readvance—Lower Gamblian Pluvial	37 ± 16
	Tazewell, Hackensack (Brandenburg) Readvance—Lower Gamblian Pluvial	23 ± 5.5
	Brady (Aurignacian, Caspian) Interstadial	18.7± 3.2
	Cary, Mankato,* Port Huron (Pomeranian) Readvance—Upper Gamblian Pluvial	14.9± 2.7
	Two Creeks (Alleröd) Interstadial	11.4± 2.2
	Valders (Fennoscandian) Readvance Makalian Pluvial	10.3± 1.8

* The date of Mankato readvance is uncertain. Some authors equate it with the Valders readvance, but recent evidence suggests that it occurred earlier.

The multiplicity of names in Table 28 is unfortunate but necessary until definite coincidence of the ice and pluvial ages can be established. The pluvials get their names from lakes in Africa. The European names for the ages and subages are given in parentheses. All other names follow the North American nomenclature that will be used in this discussion unless otherwise indicated.

As shown in the table, the Pleistocene epoch consisted of at least four glacial ages and three interglacial ages. During the former, the average world temperature was about 6° below its present value; and during the latter, temperatures averaged 3° above those of today (Fairbridge, 1961). Thus, the present climate is more interglacial than glacial.

During the ice ages, glaciation was worldwide (Hubbs, 1957). In the northern hemisphere, ice caps were most widely developed in North America, Greenland, and

Scandinavia, with the North American ice cap being much larger than that in Europe (Schwarzbach, 1963). Approximately 9 percent of the earth's surface, or 30 percent of the land surface, was ice-covered at the peak of the ice ages (Flint, 1957), with glaciation occurring as far south as the White Mountains of Arizona. Only 3 percent of the earth's surface is ice-covered today. In many regions the extent of the glaciation decreased from the first to the fourth of the ice ages, perhaps as a result of a gradual decrease of precipitation (Hubbs, 1957). This was apparently true not only in the northern hemisphere but also on the Antarctic continent (Pèivé, 1960). There are indications that the center of the glacial activity shifted eastward from Canada to eastern Europe between the first and last ice age (Pauly, 1952). In Alaska the climate was comparatively mild during the glaciations.

The ice ages as a whole were undoubtedly characterized by reduced global precipitation and evaporation; nevertheless, south of the ice sheets and on the poleward limits of what are now deserts, precipitation was apparently between two and three times greater than what occurs today (Brooks, 1951). The largest increase probably occurred in winter and was the result of the equatorward displacement of the middle latitude storm track. The presence of cold-water fossils in the equatorial and subtropical seas (Hubbs, 1957) suggests that the oceans were 5 to 10° cooler during the ice ages than during the interglacial ages (Emiliani, 1957). At the peak of glaciation, spruce forests covered Florida and Texas, reindeer and musk-oxen invaded the central United States, and walruses were found off the coast of Georgia (Dorf, 1960).

The coldest and driest of the four ice ages was the Wisconsin, with temperatures averaging about 5° lower than today in the eastern Pacific. Extremely cold conditions prevailed in North America as far south as the Mexican border. Emiliani (1957) and Hubbs (1957) have attributed the faunal extinction that occurred at this time in the United States to the very arid conditions that followed the glacial advance.

The African pluvials, which are generally assumed to be contemporary with the ice ages, were characterized by the formation of a succession of large lakes in East Africa (Brooks, 1951). The greatest of these was Lake Kamasia, which has been associated with both the Kansan and Illinoian glacials. The mean annual precipitation during its formation was probably about 180 cm, or somewhat less than twice the present average. The later Gamblian pluvial had an average rainfall of between 110 and 130 cm per year. According to Butzer (1958), the Kanjeran was the coldest and wettest pluvial in the Near East.

The interglacial ages were typically warm, dry, and ice-free. The Yarmouth was the longest and also the driest, and the Sangamon was the shortest. Evidence from the Near East (Butzer, 1958), Europe (Flint, 1957), and North America (Brooks, 1951) suggests that all three regions were warmer and drier than they are today.

During the Wisconsin ice age there was a series of glacial advances and retreats. One of the best documented of the former is the Tazewell readvance, which occurred about 23,000 years ago. Temperatures in the central Atlantic and the Caribbean and in Europe were between 7 and 12° lower than they are now, with the anomalies increasing northward and suggesting an intensified poleward temperature gradient (Öpik, 1957). The Near East was cool and wet north of 30°N and cool and dry south of 22°N (Butzer, 1958). The United States also had a cool and wet climate. Glaciation was recorded in the midwestern part of the country as far south as Iowa and Nebraska. It was during this period that Searles Lake formed in southeastern California (Flint and Gale, 1958). Martin (1958) has suggested that Greenland, the Arctic Islands, and Alaska may have been unglaciated at this time because of a lack of precipitation.

The Cary readvance, about 15,000 years ago, was characterized by a tundra-type climate in Europe (Brooks, 1951) and cool, wet weather in the western United States (Antevs, 1955). July temperatures were 7° lower and precipitation 65 percent greater than at present in New Mexico. Lake Bonneville reached its maximum height (1,450 m above sea level) at this time.

The last of the widespread periods of refrigeration was apparently the relatively weak Valders readvance, which was moist and cold in Europe (Flint, 1957), the Near East (Butzer, 1958), and the United States (Sears, 1958). All subsequent glacial advances were either local or have not been definitely established as worldwide.

The best-known and best-dated of the interstadials is the Two Creeks interstadial, which occurred about 11,400 years ago. This was a relatively mild period of glacial recession. Temperatures averaged somewhat lower than they do today over most of the northern hemisphere. Paleotemperature data indicate that a sudden warming of the Atlantic Ocean took place about 11,000 years B.P. Ewing and Donn (1956) attribute the warming to an abrupt freezing of the Arctic Ocean, suggesting that by the end of the Mankato readvance so much water was locked in glaciers that the sea level fell almost enough to cut off the Arctic from the Atlantic Ocean. As a result, the former froze and the latter warmed, since it no longer exchanged water freely with the northern seas. This warming is often considered to have marked the end of the last major glacial readvance of the Wisconsin age.

The climate of the past 10,000 years has been reviewed adequately by Brooks (1951) and is not discussed in detail here. A relatively brief sampling of climatic events for this period is given in Table 29. The outstanding features are the Cochrane glacial readvance (6800 to 5600 B.C.), the Climatic Optimum (5600 to 2500 B.C.), the Little Ice Age (A.D. 1500 to 1900), and the recent (since A.D. 1880) worldwide warming that seems to have ended between 1940 and 1950.

TABLE 29

A BRIEF CHRONOLOGY OF THE CLIMATE OF THE LAST 10,000 YEARS

Dates	Region	Climate	Sources
9000–6000 B.C....	Southern Arizona	Warm and arid.	Martin (1963)
7800–6800 B.C....	Europe	Cool and moist, becoming cool and dry by 7000 B.C. Ice sheets left Sweden in 6840 B.C.	Brooks (1951) Antevs (1955)
6800–5600 B.C....	North America, Europe	Cool and dry, with possible extinction of mammals, particularly in Arizona and New Mexico. Cochrane readvance in Alaska and southeast Canada.	Brooks (1951) Flint and Deevey (1951) Martin (1958) Sears (1958) Terasmae (1961)
5600–2500 B.C....	Both hemispheres	Warm and moist, becoming warm and dry by 3000 B.C. (Climatic Optimum). Intermittent drought in the western United States after 5500 B.C. Start of glacial retreat in the McMurdo Sound region of Antarctica about 4000 B.C. Maximum glacial retreat in Alaska near 3500 B.C.	Brooks (1951) Antevs (1955) Pèivé (1960) Karlstrom (1961) Gentilli (1961)
2500–500 B.C.....	Northern hemisphere	Generally warm and dry with periods of heavy rain (in Europe near 1300 B.C.; in the Near East about 1100 B.C. and between 850 and 800 B.C.; and in the western United States after 660 B.C.) and intense droughts (in Europe after 2200 B.C., between 1200 and 1000 B.C., and between 700 and 500 B.C.; in China from 842 to 771 B.C.; and in the western United States near 510 B.C.). Glaciation in Alaska (between 2380 and 1340 B.C. and near 600 B.C.).	Brooks (1951) Flint and Deevey (1951) Flint (1957) Butzer (1958)
500 B.C.–A.D. 0...	Europe	Cool and moist. Glacial maximum in Scandinavia and Ireland between 500 and 200 B.C.	Brooks (1951) Flint (1957)
330............	United States	Drought in the Southwest.	Antevs (1955)
600............	Alaska	Glacial advance.	Karlstrom (1961)
590–645.........	Near East, England	Severe drought in the Near East, followed by cold winters. Drought in England.	Butzer (1958)
673............	Near East	Black Sea frozen.	Butzer (1958)
800............	Mexico	Start of moist period.	Sears (1958)
800–801.........	Near East	Black Sea frozen.	Butzer (1958)
829............	Africa	Ice on the Nile.	Butzer (1958)
900–1200........	Iceland	Glacial recession (Viking period).	Schwarzbach (1963)
1000–1011.......	Africa	Ice on the Nile.	Butzer (1958)
1000–1100.......	Utah	Snowline 300 m higher than today.	Wright (1963)
1200...........	Alaska	Glacial advance.	Karlstrom (1961)
1180–1215.......	United States	Wet in the West.	Schove (1961)
1220–1290.......	United States	Drought in the West.	Schove (1961)
1276–1299.......	United States	"Great Drought" in the Southwest.	Antevs (1955)
1300–1330.......	United States	Wet in the West.	Schove (1961)
1500–1900.......	Europe, United States	Generally cool and dry. Periodic glacial advances in Europe (1541 to 1680, 1741 to 1770, and 1801 to 1890) and North America (1700 to 1750). Drought in the southwestern United States from 1573 to 1593.	Brooks (1951) Schove (1961) Schwarzbach (1963)

TABLE 29—*Continued*

Dates	Region	Climate	Sources
1880–1940.......	Both hemispheres	Increase of winter temperatures by 1.5°C. Drop of 5.2 m in the level of the Great Salt Lake. Alpine glaciation reduced by 25 percent and Arctic ice by 40 percent. Rapid glacial recession in the Patagonian Andes (1910–1920) and the Canadian Rockies (1931–1938).	Flint (1957) Heusser (1961) Mitchell (1961)
1920–1958.......	United States	25 percent decrease in mean annual precipitation in the Southwest.	Sellers (1960)
1942–1960.......	Both hemispheres	Worldwide temperature decrease and halt of glacial recession.	Mitchell (1961) Heusser (1961)

THEORIES OF CLIMATIC CHANGE

No completely acceptable explanation of climatic change has ever been presented, and it is extremely unlikely that one ever will, unless our knowledge of the earth's geological, biological, and climatological history is improved tremendously. During the past decade, it has become increasingly clear that no one process alone can explain all scales of climatic change. For example, Flint (1957) has presented a "solar-topographic" theory, which depends mainly on variations in the intensity of solar radiation and mountain building; and Panofsky (1956) prefers a theory involving the earth's orbital changes and mountain building.

In this section some of the more popular theories of climatic change are discussed. All but one are chiefly concerned with explaining the glacial epochs or ice ages, particularly the latter, and consider the intervening climate to be, more or less, the normal state of affairs. The one exception is the continental drift or pole migration theory, which has been used by several authors to account for the extreme warmth of the northern hemisphere north of 40°N during almost nine-tenths of the past 500 million years.

Observational substantiation of the continental drift theory has come from several sources. Paleomagnetic data of Howell and Martinez (1957) and Doell and Cox (1961) indicate that the magnetic poles have shifted relative to the continents from near the present equator and 0 and 180°W at the start of the Cambrian period to approximately their present positions at the end of the Eocene epoch. Similar results were obtained earlier by Runcorn (1956). (See also Green [1961].) The approximate geomagnetic latitudes of various regions in past geological periods are listed in Table 30. These figures are based on results given by Opdyke (1962).

The pole positions as estimated from paleomagnetic data for different continents usually do not agree. For example, measurements in Africa and Australia put the Carboniferous South Pole in South Africa and the Azores, respectively. The data become compatible only if the position of the continents relative to one another has changed in the geological past. Runcorn (1962) concludes that, until the Triassic, Europe and North America were about 30° closer together than they are today.

Recent astronomical evidence of Munk and Markowitz (1960) has definitely established that Greenland has moved about 9 m toward the geographic North Pole during the first half of the twentieth century. This is a relatively fast rate of movement, corresponding to about 91,400 km in 500 million years.

TABLE 30

GEOMAGNETIC LATITUDE OF VARIOUS REGIONS IN PAST GEOLOGICAL PERIODS

(Based on paleomagnetic data given by Opdyke, 1962)

Period	Central United States	Northern Canada	Southern Greenland	France	Australia	North Africa
Cambrian	5	19	....	24	6	....
Ordovician	2	16	....	64	....	....
Silurian	14	10	....	4	12	....
Devonian	9	36	....	3	10	....
Carboniferous	1	28	....	1	67	51
Permian	10	23	19	3	56	16
Triassic	9	44	47	22	70	29
Jurassic	34	65	....	28	59	16
Cretaceous	....	....	....	44	....	0
Quaternary	54	90	71	44	46	18

The continental drift theory has many persuasive features and some drawbacks. Perhaps the most serious of the latter is the necessary assumption that the magnetic poles have always been near the geographic poles. According to Öpik (1958*b*), this assumption is absolutely unfounded and the magnetic poles are as likely to be close to the equator as near the poles of rotation. On the other hand, Schwarzbach (1963) claims that there are theoretical grounds for believing that the pole of rotation and the magnetic pole are causally related, since the terrestrial magnetic field probably arises from currents within the earth's mantle and, therefore, should be symmetrical with respect to the axis of rotation.

The continental drift theory is not only verified by the direct observation that continents do drift, but the apparent locations of the poles in the geologic past can be used to explain the distribution of climates without requiring any net heating or cooling of the earth's surface, except that associated with mountain building and glaciation. On the basis of paleomagnetic data, Runcorn (1956) and Howell and Martinez

(1957) believe the poles were near the equator off the west coast of Africa and in the Gilbert Islands of the Pacific Ocean during the Pre-Cambrian glacial period, which was most pronounced in South Africa, India, China, and Australia. In the Carbo-Permian glacial period, when the poles were probably in South Africa and in the southeastern North Pacific near Hawaii, traces of ice have been found only in India in the northern hemisphere. On the other hand, vast ice sheets covered the sub-tropical regions of Africa, Australia, and South America.

The warmth of the region north of 40°N during most of the past 500 million years can be explained very satisfactorily by the continental drift theory. According to Brooks (1951), the average temperature in this region in the late Cambrian was about 15°C. But during this period, the equator probably passed through the eastern United States, Greenland, western Asia, and possibly Antarctica. Thus the region now north of 40°N was then within the latitude zone between 50°N and 50°S, which today has an average sea level temperature of about 20°C. This value agrees well with that of Brooks if a mean height of 600 m is assumed for the continental regions north of 40°N during the Cambrian period.

Most proponents of the continental drift theory visualize a very slow apparent movement of the poles, certainly no faster than about 50 cm per year; however, there are some scientists who believe that shifting of the earth's crust may sometimes be a relatively quick process, occurring fast enough to produce the glacial epochs and to freeze almost instantly any animals unlucky enough to be caught on a continent that is suddenly displaced poleward in one cataclysmic heave (Hapgood, 1959; Sanderson, 1960). In order to make this shifting-crust theory physically possible, the continents are considered to be "a kind of floating islands in the pitch-like viscous ocean of the underlying magma, apt to change their position with respect to each other and the axis of rotation" (Öpik, 1957). Invariably those who favor this theory refer to the sudden extinction of mammoths, giant beavers and sloths, and saber-toothed tigers about 10,000 years ago and point to the presence of well-preserved remains of mammoths in Siberia and northern Alaska as proof that major climatic changes have occurred almost instantly over large sections of the earth. They believe that these changes could occur only by rapid shifts of the earth's crust.

The reason that continents drift remains to be explained whether the shifts be slow or fast. Plausible theories have been presented by Munk and Markowitz (1960), and by Runcorn (1962) and Vening Meinesz (1962).

Munk and Markowitz (1960) attribute the current polar shift to the melting of the Greenland ice cap. They reason that since the earth rotates about an axis which passes through the North and South poles, a centrifugal or outward force is exerted on all particles on its surface. This force is directly proportional to the distance of the particles from the axis of rotation and hence is greater, for example, on a tall mountain than in a deep valley at the same latitude. It is also greatest at the equator and least (zero) at the poles. If an ice cap forms at some point on the earth's surface other than at the equator, an excess of centrifugal force about the earth's axis is created, tending to force the ice cap outward, that is, equatorward. If the earth's crust is assumed to be semirigid, so that there is little movement of one continent relative to any other, a shift of an ice cap equatorward implies a shift of all the features of the earth and an apparent movement of the poles.

Similar reasoning is, of course, applicable to a melting ice cap, except that in this case the centrifugal force exerted by the cap decreases as the ice melts, and this causes a slow drift poleward. This is what Munk and Markowitz believe is happening today.

Presumably the building or weathering of mountains might also cause the continents to drift; however, Hapgood (1959) suggests that orogenesis is more a result than a cause of pole migration. He believes that if a section of the earth's crust is moved poleward, it will be slightly compressed and must fold, producing mountains. On the other hand, a section moved equatorward will be stretched, forming volcanic zones.

The Munk-Markowitz theory very effectively explains why, during nine-tenths of the geologic past, most of the major continents were on or near the equator; however, it cannot alone explain either the glacial epochs or the ice ages, since regions of orogenesis and glaciation should migrate away from and not toward the poles.

The Runcorn-Vening Meinesz theory is of geophysical origin. Runcorn (1962) suggests that the flow of heat out of the earth's core produces thermal convective currents moving at speeds of about 10 cm year^{-1} within the earth's mantle. The primary result of the convection is to determine the location of major topographical features of the earth by causing the continents to move toward regions where the currents are descending. Vening Meinesz (1962) presents several arguments showing that such convection actually does exist. The currents are cellular, and the number of cells and the velocity of the motion depend primarily on the composition of the mantle, the temperature difference between ascending and descending currents, and the ratio of the core radius to the earth's radius.

The present core radius, 0.55 of the earth's radius, favors the existence of five convective cells with continents centered near the South Pole, 18°S, and 54°N, almost as observed (see Table 1). Following Urey (1952), Runcorn (1962) assumes that the earth had a cold origin and that its core began to grow about 3,000 million years ago and is still growing very slowly today. He then concludes that before the continental drift—deduced from the geological record—took place, the core radius was close to 0.50 of the earth's radius, favoring the existence of four convective cells with continents centered at the poles and the equator.

The Runcorn-Vening Meinesz theory of continental drift, although still speculative, is probably more plausible than the Munk-Markowitz theory; however, it too fails to explain the relatively short glacial epochs. Perhaps the two suggested mechanisms work together, the equilibrium positions of the continents being determined by convection within the mantle and departures therefrom by the excess centrifugal force exerted on regions of orogenesis and glaciation.

The remainder of this section deals with those theories of climatic change whose main purpose is to account for the glacial epochs and ice ages. More than fifty such theories have been presented in the literature. It would be a hopeless task to discuss all of them adequately in the space available; therefore, consideration is given only to those that have received popular support within the past decade—the theories of mountain building, volcanism, variations of sea level, carbon dioxide, variations in the intensity of solar radiation, and variations of the earth's orbit around the sun.

The value of each theory in explaining the ice ages rests almost entirely upon its ability to account for what is known about the climate of the Pleistocene epoch. Basically, each must show why, from an interglacial to a glacial age, the poleward temperature gradient and the atmospheric circulation intensify; the surface temperature decreases at all latitudes of both hemispheres, with the greatest drops occurring near the poles; vast ice sheets form at high latitudes; and precipitation in the subtropics increases.

Mountain building.—Mountain building is a popular theory of climatic change, particularly among geologists, who have found excellent correlations between the general periods of orogenesis and subsequent glaciation. Both the Carbo-Permian and Quaternary glacial epochs are known to have been preceded by periods of extensive mountain building. That mountains are necessary for glaciation is now generally accepted, but it is also true that some major orogeneses have been accompanied by little if any glaciation. The most notable example of this is the Caledonian orogenesis, which took place in northwestern Europe 370 to 450 million years ago. Although glaciation did occur in the Devonian period, it was local and restricted mainly to South America and South Africa. If, however, Howell and Martinez' positions for the poles during this period are correct (35°N, 155°W, and 1,600 km southeast of the southern coast of Brazil), then the equator of that day passed through northwest Europe. If this was the case, even with the increased elevation, the climate may have been too warm for glaciers to form.

Another commonly quoted objection to mountain building as an important factor in producing glaciation is that active mountain building has preceded the glacial epochs by several million years (Öpik, 1957). The geologists' answer to this criticism is that ice sheets develop most easily from glaciers formed in ancient mountain ranges that have been eroded into peaks and valleys.

Some geophysicists contend that orogenesis is always occurring somewhere on the earth (Panofsky, 1956). This is perhaps the most damaging argument against the mountain building hypothesis because, if this is true, mountain building would have to be considered at best a necessary but not sufficient condition for glaciation. In any case, because of the time scale involved, the theory cannot explain the ice ages.

Volcanism.—Volcanism encourages glaciation by discharging into the atmosphere very small dust particles that are large enough to scatter the incoming shortwave solar radiation—part of which is returned to space unaltered—but too small to have appreciable effect on the longwave terrestrial radiation. It is doubtful, however, whether even the most intense volcanic activity could reduce the earth's temperature by much more than 1°; also, all but the most minute particles would settle to the surface within a few years. Nevertheless, in those cases where mountain building is not enough, volcanism may act as a trigger for glaciation, which once started may be self-perpetuating. Since volcanism usually is most intense at the same time that orogenesis is at a maximum, the theory cannot account for the lag of several million years before the onset of glaciation.

Although volcanism should probably be rejected as a major cause of glacial epochs or ice ages, it is still favored by some scientists. Wexler (1960) believes that since continents react much more rapidly than ocean areas to any change in the intensity of incoming radiation

a period of sustained volcanic activity of the kind which injects fine ash high into the atmosphere, where it floats for many years, would cause the land areas of the earth to become much colder than the ocean areas. The increased temperature contrast between land and ocean in winter would result in increased storminess and precipitation—a winter condition which is favorable for the formation and nourishment of glaciers. In summer, the presence of dust in the high atmosphere, and possible additional clouds that would result from the nucleating action of the dust, would tend to keep the earth cool, and thereby preserve the snow and ice from one winter to the next. It is thus possible to account for the glacial ages on the basis of increased turbidity in the atmosphere.

Mitchell (1961) suggests that the short-term transient irregularities in what otherwise might appear to be highly regular, long-term variations of global mean temperature are the result of volcanism. He also indicates that the relatively rapid rate of cooling of the northern hemisphere since 1942 is due to a marked increase in major volcanic activity.

Variations of sea level.—Ewing and Donn (1956) accept the continental drift theory to explain the long-period climatic changes, but they attribute the Pleistocene ice ages to variations in the level of the Arctic and Atlantic oceans. Their theory is based on evidence of sudden warming throughout the North Atlantic Ocean about 11,000 years ago at the end of the Mankato readvance (Ericson *et al.*, 1956). They suggest that before the peak of the ice age and before a maximum amount of the earth's water was locked in glaciers, a free interchange of water took place between the Arctic and Atlantic oceans across a shallow sill between Norway and Greenland. As the glaciers grew they continued to take moisture from the surrounding ice-free Arctic Ocean, until, about 11,000 years ago, the sea level fell below the level of the sill. At this time the Arctic was presumably cut off from the warmer water to the south and froze, thus depriving the glaciers of their main source of moisture and ending the ice age. In the meantime the Atlantic Ocean warmed, since it no longer exchanged water freely with the relatively cold Arctic Ocean.

It is difficult to discount this interesting theory of the Pleistocene ice ages. There are many observations that it can explain, such as the existence of the greatest ice sheet development in the eastern half of the United States and a temperate Alaska. But, on the other hand, there are some facts that it either cannot explain or that have to be adjusted considerably to fit into the over-all scheme. For example, most authorities agree that geological evidence points to an open Arctic Ocean during the climatic optimum 7,600 to 4,500 years ago. However, Ewing and Donn (1956) maintain that this evidence pertains to the open Arctic they postulate for the Mankato readvance about 15,000 years ago. Also, their theory cannot explain the present warming of the Atlantic Ocean, which is taking place in spite of the melting of arctic ice and a rise in sea level.

On a much larger scale than visualized by Ewing and Donn, variations of sea level must have a significant effect on climate. A 25-percent reduction in the area of exposed land surface, due to an increase of sea level, would lead to an increase of slightly less than 3 percent in the worldwide average evaporation and precipitation, if present values over land and water are any indication. Although the percentage increase is small, the released latent heat would warm the atmosphere by more than 7° and probably be sufficient to produce a warm interglacial period. Such conditions probably existed near the start of the Pliocene epoch, when, according to Zeuner (1959), sea level was 380 m above the present.

According to Fairbridge (1961), mean sea level has fallen about 100 m since the beginning of the Pleistocene epoch (Fig. 47). Superimposed on the over-all trend are periodic rises and falls associated with glacial retreats and advances, respectively. The latest rise began about 17,000 years ago and continued until about 4000 B.C. Since that time relatively minor fluctuations have occurred, with a variable rise of 2 m indicated for the past 2,000 years.

Although the gradual decrease of sea level since at least the beginning of the Pleistocene epoch can be attributed to the accumulative storage of water in the ice caps from one ice age to the next, Zeuner (1959) believes it more likely that the drop

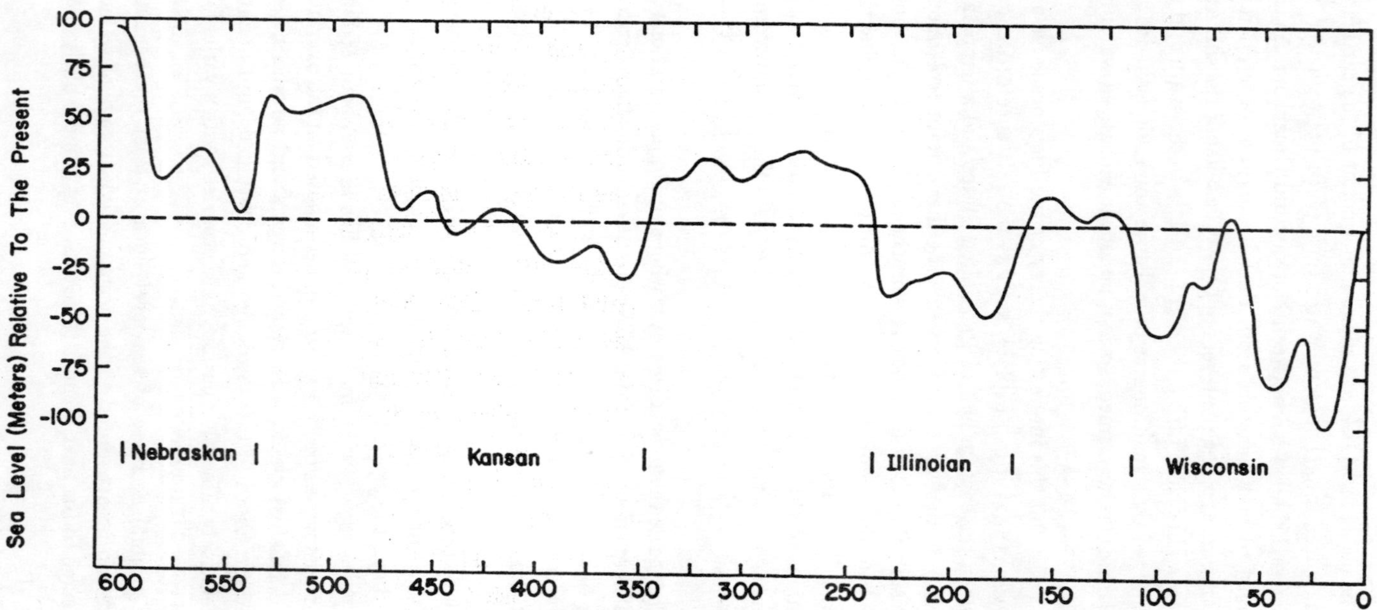

FIG. 47.—Pleistocene variations in sea level (meters) relative to the present. Based on observations of interglacial "raised beaches" and on correlation with astronomical variations in the earth's orbit around the sun. From Fairbridge (1961).

is caused by a sinking of the bottom of the sea. Isostatic adjustment of the earth's crust then requires that the continental areas must be rising. This conclusion is also strengthened by an observed drop in sea level through most of the Tertiary period, when there were no glaciers.

The progressive drop of sea level through the ice ages of the Pleistocene is in line with an accompanying decrease in precipitation, as suggested by Hubbs (1957), and with the belief that the Wisconsin was the coldest and driest of the four ice ages. This theory of eustatism, as it is called by Zeuner (1959), cannot, by itself, explain the interglacial ages or the periodic advances and retreats that have occurred during the ice ages.

Carbon dioxide.—Some authors have attempted to relate climatic change to variations in the carbon dioxide content of the atmosphere. Carbon dioxide is an excellent absorber of longwave terrestrial radiation, particularly in that part of the electromagnetic spectrum lying between 12 and 18 μ. Hence, a small increase in its concentration would presumably raise the temperature of the atmosphere by reducing the amount of terrestrial radiation lost to space.

One of the most often quoted objections to this theory is that, since water vapor absorbs strongly in the same spectral regions as carbon dioxide, the carbon dioxide can have little influence on the terrestrial flux; however, Plass (1956, 1957) points out that, because the concentration of water vapor decreases rapidly with height in the atmosphere and because its absorption lines are distributed randomly with respect to those of carbon dioxide and vice versa, the effect of water vapor is much smaller than has generally been believed.

Another objection is that, since even a small amount of carbon dioxide almost completely absorbs the terrestrial radiation in the 15-μ band, any decrease or increase of the concentration may not have much effect on the temperature. Plass answers this objection by showing that, with normal cloud cover, the average surface temperature must increase by 2.5° or decrease by 2.7° when the carbon dioxide amount in the atmosphere is doubled or halved, respectively. He assumes that carbon dioxide variations occur independently of fluctuations in other factors that influence the radiation balance and that the temperature changes by equal amounts over land and water.

The development of the carbon dioxide theory was encouraged by a simultaneous increase in the temperature and carbon dioxide content of the atmosphere during the first part of this century. Since 1942 the carbon dioxide content, however, has continued to rise at a rate of about 0.2 percent per year (Callendar, 1958; Eichhorn, 1963), while the average world temperature has fallen slightly (Mitchell, 1961). Further, the greatest warming between 1900 and 1940 occurred in Scandinavia, far from the major industrial centers where carbon dioxide might be expected to accumulate as a result of the combustion of fossil fuel.

Using the carbon dioxide theory, Plass (1957) attempts to explain the lag of several million years between orogenesis and glaciation. Mountain building itself releases large amounts of carbon dioxide from the earth's interior through volcanic vents and hot springs, thus indirectly warming the atmosphere. This excess carbon dioxide plus possibly half of that originally present must be withdrawn before the atmosphere can be cooled enough for glaciation to start. The process suggested for accomplishing this withdrawal is the weathering of igneous rock exposed during the period of orogenesis. This is a very slow process, however, and one which must continue for several million years before the atmosphere can cool enough for glaciers to form. Plass has used the same argument to explain orogenesis without glaciation. In this case the amount of carbon dioxide released from the earth's interior is so large that weathering can never reduce the atmospheric content to a sufficiently low level to produce glaciation.

The ice ages of a given glacial epoch are explained by using the concept of carbon dioxide equilibrium between the oceans and the atmosphere. Once the atmosphere is cooled enough for an ice age to start, the oceans begin to lose water to glaciers, which can permanently hold only relatively small amounts of carbonates. Thus, the carbon dioxide concentration of the oceans increases, offsetting the original balance that existed with the atmosphere. As a result, during the thousands of years required for the glaciers to grow into continental ice sheets, and probably for a long period thereafter, carbon dioxide is diffused into the atmosphere. Finally, a new equilibrium condition is established, usually several tens of thousands of years after the initial start of the ice age. But the increased carbon dioxide concentration in the atmosphere leads to warming, melting, enlargement of the oceans, and an interglacial age. With the oceans back to their original volume, equilibrium is again destroyed and carbon dioxide must once more be removed from the atmosphere. This process leads to another ice age and must continue until the total carbon dioxide content of the atmosphere-ocean system is increased enough to keep the air temperature above the value required for glaciation. Plass does not present a convincing argument on how this increase can occur.

The carbon dioxide theory is weakest quantitatively. In order for the atmosphere to cool sufficiently for an ice age to begin, its carbon dioxide content must be reduced by at least 50 percent of the present value. Although this is equivalent to a decrease of only 7 percent in the total carbon dioxide content of the atmosphere-ocean system, it may still be larger than could occur by natural processes. It is interesting to note, however, that if the combustion of fossil fuels continues to increase at its current rate, by the middle of the next century the carbon dioxide content of the atmosphere will be four to ten times larger than at present and the surface temperature of the earth will have risen 7 to 12°. A temperature increase even half this great would be disastrous. It would melt the ice caps, inundate many densely settled coastal areas, annihilate many life forms in equatorial regions, and destroy commercial fisheries (Eichhorn, 1963).

Plass (1961) also considers the effects of ozone and water vapor on climate, concluding, first, that a lowering of the atmospheric level of the ozone maximum from 28 to 10 km will increase the surface temperature by 2.1° and, second, that temperature changes produced by variations in ozone and carbon dioxide are augmented by water vapor, whose average concentration normally increases with temperature. He fails, however, to consider the indirect cooling of the atmosphere after the ice caps have formed. If this cooling were taken into account, the carbon dioxide content of the atmosphere might have to increase by a factor of three or four before a glacial retreat could begin.

Kaplan (1960) has reworked Plass's approximate calculations, taking into account the detailed structure of the 15-μ carbon dioxide band and its dependence on the thermal structure of the atmosphere. He concludes that, even neglecting the shielding effect of water vapor, which is far from negligible, halving the carbon dioxide of the atmosphere would result in a temperature decrease of only 1.8°. Thus, it appears that for carbon dioxide to play a major role in climatic change, its atmospheric concentration must vary by one order of magnitude from an ice age to an interglacial age. There is currently no evidence which suggests that such large variations actually occur.

Variations in the intensity of solar radiation.—All theories of climatic change discussed so far have been based on variations of either terrestrial or atmospheric factors. It has been tacitly assumed that the earth has received a constant amount of radiation from external sources, primarily the sun, since at least the Cambrian period. There are many scientists who question this assumption and present their own theories based on fluctuations in the intensity of solar radiation as the prime factor in producing climatic changes. There are two schools of thought. The first states that ice ages are associated with a decrease of solar radiation or with a relative decrease of temperature. The second claims that ice ages are associated with increased solar radiation or with a relative increase of precipitation.

The "glaciation-with-a-cool-sun" school consists mainly of geologists and astronomers. They believe that ice caps can develop and grow only when the intensity of solar radiation has decreased below some critical value. Observations from the tropics confirm this belief to the extent that glaciation has been associated with a decrease of temperature at *all* latitudes.

The increase in precipitation during the ice ages, in spite of a cooler atmosphere, can be explained in at least two ways. First, heavy precipitation in Africa and other subtropical regions during the Pleistocene glaciations could have resulted from the solar-induced lower temperatures, which caused reduced evaporation, and the expansion of the ice sheets. The middle latitude storm track usually lies in the zone of greatest poleward temperature gradient, which in winter is normally between the snow-covered ground to the north and the exposed surface to the south. During ice ages or unusually severe winters, the storm track should lie much farther south over the continents than it does today, thus favoring heavy rains, particularly in winter, in subtropical regions. Actually neither of these factors—decreased temperatures or an expanding ice sheet—would necessarily lead to an increase in the worldwide precipitation. But the wettest regions would shift equatorward with the ice sheet, and precipitation lost by evaporation would decrease.

A second explanation for the increased precipitation is the same used by Wexler (1960) in his defense of the volcanism theory. It is based essentially on the time required for the oceans to cool following a decrease of solar radiation. This period should be approximately the same as the average period of circulation for the deep waters of the oceans, which Plass (1961) estimates as being several thousand years. On the other hand, the continents would cool almost immediately and become much colder than the oceans in winter, especially at high latitudes. In North America and Asia, this would lead to an almost continuous generation of very cold, continental, polar high pressure cells that, one after another, would sweep down into middle latitudes, producing blizzards and cold waves worse than any modern man has ever experienced. In other areas, the passage of warm moist air from the oceans over the cold land would lead to long periods of fog and drizzle. The warmth of the oceans is essential to this argument, for only then can the relatively cool and unstable atmosphere pick up enough moisture while passing over the warm water to produce heavy snowfall on land.

The glaciation-with-a-cool-sun theory and all the other theories based upon fluctuations of solar radiation have the most difficulty trying to explain how the intensity of solar radiation can vary. Öpik (1950, 1953, 1958*a*, 1958*b*), in presenting a solar explanation for the complete spectrum of climatic change, suggests that the solar constant has been gradually increasing during the last 4,500 million years as a result of a slow decrease in the hydrogen content of the sun's hot core. The sun produces heat by the nuclear process of fusion, converting hydrogen into helium within the core. As the available hydrogen is used up, the amount of heat produced by the core decreases. Then, in order that energy can be radiated away as fast as it is produced, the sun must contract. The net result is a hotter sun, the heat deficit apparently being more than equaled by the release of gravitational energy.

Öpik concludes that 4,500 million years ago 33 percent of the sun's core was hydrogen and the mean temperature of the earth was −17°C, resulting in a "permanent" ice age. Today the hydrogen content of the sun's core is estimated at 10 percent and the temperature of the earth at 13°C. In one billion years, the hydrogen content will presumably be lowered to 1.25 percent and the earth's temperature raised to 34°C.

Öpik attributes the glacial epochs of the past 500 million years to temporary reversions of the sun to an earlier state, with the high hydrogen content of the core being induced by an expansion of the core to include a heat barrier composed of carbon, nitrogen, oxygen, neon, magnesium, silicon, and iron. This barrier is assumed to form when hydrogen is slowly diffused into the core from the surrounding mantle. As for the ice ages within a glacial epoch, Öpik states:

> These fluctuations seem to be worldwide and have been most difficult to understand. My own guess is that they may represent a kind of "flickering" of the disturbance in the sun—like a candle flame blown by the wind. They may be connected with the irregular mixing of matter in and around the core of the sun during the disturbance, or with allied events at the surface.

Perhaps the main advantage of Öpik's theory for the glacial epochs is that it permits a recurrence of the glacial epochs at 250 million year intervals, since a period of about this length is required to build the "metal barrier" around the sun's core that starts the disturbance. This process must continue as long as hydrogen diffuses into the core.

The main disadvantage of the theory is that most of it is based too exclusively on speculation. Observations have been taken for too short a period to establish definitely that the solar constant varies significantly, although Johnson and Iriarte (1959) measured an increase of about 2 percent in the solar constant between 1953 and 1958. Mitchell (1961) correlates this increase with a simultaneous increase in sunspot activity.

The concept of a slowly contracting and warming sun does not seem to be generally accepted among astronomers. According to Menzel (1959) and Gamow (1964), when the hydrogen within the central core of the sun is exhausted, the core will contract and the temperature at its outer boundary will reach 20 million degrees. Hydrogen will then begin to burn in a thin shell around the core. With time, this shell will advance outward toward the sun's surface, causing the sun to expand to many times its present diameter and to shine with many times its present brightness. A similar conclusion is reached theoretically by Schwarzschild (1958).

Some meteorologists, notably Simpson (1934, 1957, 1959), Bell (1953), and Willett (1953), reject the cool sun theory on the grounds that it does not adequately explain how an intensified poleward temperature gradient and storminess, as well as an intensified poleward transport of heat and water vapor in the lower middle latitudes, can be maintained by a tropical zone which is cooler, is receiving less heat, and is evaporating less water than at the present time. Simpson, therefore, presents his own theory, which is based upon cyclic variations in the magnitude of the solar constant with a period of about 380,000 years. He believes that a slight increase of the solar constant above its present value would lead to a stronger poleward temperature gradient, since the heating would be greatest near the equator. The increased gradient would strengthen the atmospheric circulation, which in turn would increase evaporation (above the increase due to the heating alone) and the moisture content of the air. The result would be heavier precipitation in polar regions and growth of the ice caps.

There are several features of Simpson's theory that make it hard to accept. First, in order to fit the observations, Simpson assumes that when the solar constant increases beyond a certain limit, the increased heating abruptly reverses the process of glaciation. This would lead to an extremely wet interglacial period with a strong glacial pattern of the atmospheric circulation, an occurrence for which there is no geological evidence. Second, observations indicate that the tropical Atlantic Ocean was at least 6° cooler during the ice ages than it is now. Warmer temperatures would be required if glaciation were associated with increased solar radiation. Finally, unlike the glaciation-with-a-cool-sun theory, this theory cannot explain the reoccurrence of glacial epochs at 250 million year intervals.

It should also be mentioned that the major factor in creating the strong zonal circulation of the ice ages is not excessive heating in tropical latitudes but rather the presence of an ice sheet in high latitudes, which forms most readily in a moderate climate with mild summers.

In order to avoid some of the difficulties in Simpson's theory without rejecting it altogether, Willett (1953, 1961, 1964) proposes that climatic cycles are caused, not by variations in the effective radiation of the sun, but rather by irregular solar activity of the type that accompanies sunspot disturbances. For the physical causative agency that acts directly on the upper atmosphere, he distinguishes between short wavelength ultraviolet radiation and emissions of charged particles, both of which are directly measurable only from satellites outside the earth's atmosphere.

A significant part of the radiational fluctuations of the sun occurs in the ultraviolet portion of the electromagnetic spectrum. Since only 9 percent of the total solar radiation is in this region, however, these fluctuations can produce only negligible changes in the solar constant. Increased emissions of ultraviolet solar radiation are associated mainly with variations in the intensity of the hydrogen Lyman-alpha line at 0.1216μ in plage areas of the sun (Rense, 1961). Ultraviolet radiation is very closely related to sunspot activity, for which data are available as early as 1750. The region of its maximum intensity in the earth's atmosphere moves north and south with the sun in tropical latitudes.

The emission of charged particles (protons) from the sun produces intense electric currents which enter the earth's atmosphere at about 6,000 km sec^{-1}. Because of these high velocities, relative to those of average molecules in the upper atmosphere, each proton represents a great amount of energy. Corpuscular radiation of this sort is associated mainly with magnetic storms and auroral activity, which occur during periods when the radiation reaches the earth in large amounts. Both phenomena usually occur in high latitudes, auroral activity reaching a maximum at about 26 degrees of latitude from the geomagnetic poles. The intensity of corpuscular radiation cannot be correlated well with sunspot activity. Although the two tend to occur together, there are many observations of corpuscular radiation that are not associated with sunspots. It has also been noted that corpuscular radiation tends to increase in intensity for several years after the sunspot number has reached its maximum.

Willett believes that irregular solar activity can explain the entire complex of climatic cycles that have occurred since the start of the Pleistocene epoch. He points out that no over-all heating of the atmosphere is required to produce these cycles. The solar activity, which does not affect all parts of the earth equally, is assumed to redistribute the heat and cold sources of the lower atmosphere in such a way that the circulation pattern is modified. An increase in ultraviolet radiation is accompanied by above-normal heating in the tropical stratosphere and the ionosphere. Craig (1952) shows that this high-level warming leads to a strengthening of the zonal circulation pattern and, hence, to a pattern with definite glacial characteristics. On the other hand, the direct heating of the polar stratosphere by excessive solar corpuscular radiation results in "highly chaotic periods of climatic stress, with markedly contrasting extremes of storminess, rainfall, and temperature in middle latitudes, accompanied by anticyclonic blocking of the circulation pattern and the widespread occurrence of hot dry summers" (Willett, 1953). These periods correspond most closely to Simpson's warm, wet interglacials. The temperate, dry interglacials are generally periods when both ultraviolet and corpuscular radiations are low.

Since sunspot activity is related to both ultraviolet and corpuscular radiation, its connection with climatic change is somewhat confused. On the basis of 200 years of data, Willett (1964) notes that sunspot activity, as indicated by the Zurich relative sunspot number (Fig. 48), has cyclic variations of 80 and 11 years, with the latter being composed of alternating maxima of high and moderate intensities (the double sunspot cycle).

The 80-year cycle is characterized by a gradual increase in the intensity of the sunspot maxima, from a very low level during the first quarter of the cycle to a very high level during the last quarter. Willett (1964) associates periods of low sunspot activity, which occurred from 1800 to 1820 and from 1880 to 1900, with increasing zonal circulation in low latitudes and glacial advance. Glacial recession and climatic warming occur primarily during the third quarter of the 80-year cycle. At this time, the climate is warm and wet poleward of 50° and warm and dry equatorward of 50°. The last quarter of the cycle, with intensified sunspot maxima, is a period of climatic stress, with a strong meridional circulation, warm dry summers in the continental interiors, and possibly glaciation on the windward side of the continents in high middle latitudes. Such a period ended with the sunspot maximum of 1957.

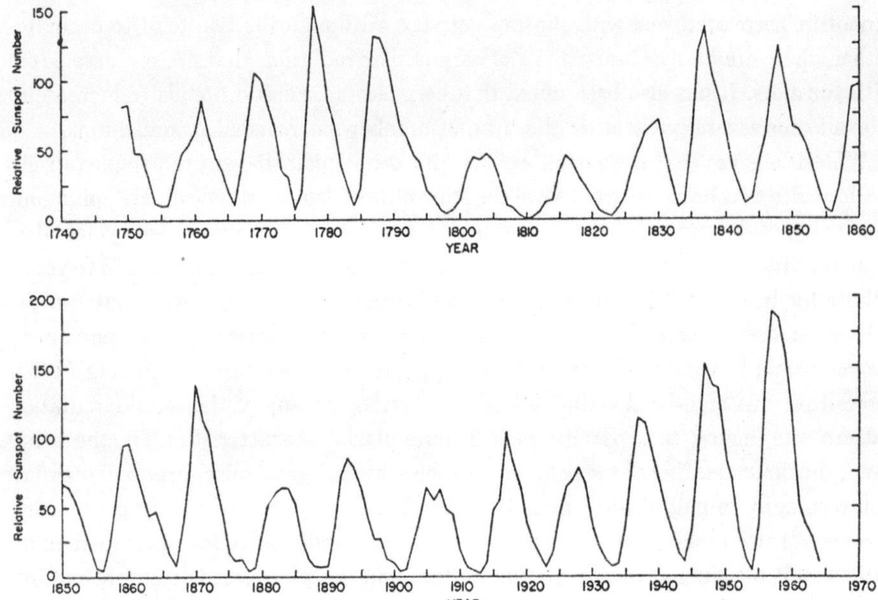

FIG. 48.—Mean annual sunspot number for 1749 to 1964. Based on data given by Clayton and Clayton (1944, 1947), Conover (1959), and Waldmeier (1956–1964).

The double sunspot cycle is less regular than the 80-year cycle, as is apparent from Figure 48, and the climatic changes associated with it are not always obvious. The sequence visualized is one of glacial advance between a sunspot minimum and the following minor maximum, of glacial recession between the minor maximum and the following minimum, and of climatic stress between the minimum and the following major maximum.

On the basis of his observations, Willett (1964) speculates that the climatic changes during the 80-year cycle are the result of significant long-term variations of the solar intensity in the ultraviolet and visible portions of the electromagnetic spectrum. He tentatively relates the double sunspot cycle, which is most apparent in winter at high latitudes, to variations in atmospheric ozone caused perhaps by changes in solar corpuscular radiation (Willett, 1962, 1965). The latter correlation is considered extremely dubious by London and Haurwitz (1963).

Willett's theory has been at least partly accepted by a large number of meteorologists, geologists, and astronomers. It has the advantage over most of those proposed in that it has been verified on a time scale of days (Craig, 1952; Shapiro, 1958, 1959; MacDonald and Roberts, 1960). That is, sudden increases in the solar ultraviolet emission have been followed after a few days by a shift of the circulation toward a glacial pattern; and intensified corpuscular radiation, usually indicated by magnetic storms in the earth's atmosphere, has often been followed by a breakdown of the zonal circulation into a cellular form typical of a period of climatic stress. These reactions do not always occur, however, and often the statistical significance is marginal. Obviously, then, there must be other factors involved that are probably not associated directly with the solar disturbance.

Since there are indications that ultraviolet and corpuscular radiations affect the atmospheric circulation over a period of days, it is natural that Willett should suggest that these effects be extended to the ice ages and perhaps also to the glacial epochs. But this assumption is not necessarily valid, although if it were, the problem of climatic change would be simplified enormously. Until some physical process that can keep the sun in a disturbed state for tens of thousands of years is described, the variable sun theory cannot be accepted as the principle cause of the ice ages.

Furthermore, there is still no good explanation of how relatively small amounts of solar energy absorbed high in the atmosphere, probably above 50 km, can affect large-scale wind motions in the troposphere. The amount of energy generally assumed to be associated with a magnetic storm or aurora, about 1.2×10^{15} cal, is several orders of magnitude smaller than the kinetic energy difference between small and large atmospheric wind disturbances (MacDonald and Roberts, 1960).

Some investigators believe that heating at the top of the atmosphere produces ascending motions in the stratosphere that, in turn, make the lower atmosphere more unstable or aid in the vertical exchange of momentum. Others, like MacDonald and Roberts (1960), suggest that "atmospheric nuclei are produced or rendered effective as freezing nuclei by an increase in the bremsstrahlung X-ray radiation that penetrates to stratospheric levels as a secondary effect of aurora-producing solar corpuscles." The increase of freezing nuclei is assumed to affect the condensation and sublimation processes, thereby modifying the large-scale circulation. Some support for this theory is found in the partial verification of the hypothesis proposed by E. G. Bowen that the passage of the earth through meteoritic showers increases the effective freezing nuclei count and precipitation amounts (Brier, 1961). The process discussed by MacDonald and Roberts should occur during periods of climatic stress, when, according to Willett's theory, corpuscular radiation is at a maximum and precipitation is extremely variable but on the whole above the average for an ice age, except in lower middle latitudes and the subtropics.

Before we conclude this review of Willett's theory, the sunspot number must be discussed as an index of climatic change. Because sunspots are related to both ultraviolet and corpuscular radiations from the sun and because these produce essentially different effects on the atmosphere, it is difficult to anticipate how or why the climate should react to variations of sunspot activity. Willett's hypothesis based on the 80-year and double sunspot cycles is, therefore, primarily statistical, being founded largely on correlations between the sunspot number and various indexes of the atmospheric circulation. As such, it is subject to the common criticism that if enough correlations are computed, a few will falsely indicate statistical significance, particularly if there exists no physical theory to indicate what kind of relation to expect.

The sunspot activity cycles stressed by Willett are also not fully established. Although the 80-year cycle was obtained from a 200-year record of sunspot numbers, this record is statistically homogeneous only since 1850, when the index was first defined as it is today (the number of spots plus ten times the number of sunspot groups). The double sunspot cycle, with alternating high and low maxima of the 11-year cycle, is immediately obvious from the record only for the period between 1830 and 1937 and may therefore represent a transitory phenomenon on the sun which will never be repeated.

Recent studies by Brier and Bradley (1964), Bigg and Miles (1964), and Lund (1964, 1965) have almost conclusively established a significant lunar influence on weather. They find that below-average sunshine and above-average freezing nuclei concentrations and precipitation tend to occur between the new moon and the first quarter and between the full moon and the last quarter. Above-average sunshine and below-average freezing nuclei concentrations and precipitation occur between the first quarter and the full moon and between the last quarter and the new moon. These results apply only to the United States and Australia and may not be the same elsewhere. In the United States, the relationship seems best established in the Midwest and during periods of low sunspot activity.

On a much longer time scale, Pettersson (1914) and Karlstrom (1961) relate climatic changes to variations in tidal-force intensity, which reaches a maximum about every 1,700 years. At such times, the sun, earth, and moon are in line with the earth at perihelion and with the moon at perigee. Furthermore, the orbits of the sun, earth, and moon are almost in the same plane once every 567 years (1,700/3 years). Karlstrom correlates periods of glacial advance in Alaska with alternate 1,700-year cycles.

Variations in the earth's orbit around the sun.—One of the most discussed, debated, accepted, and rejected theories of climatic change is the one based on variations in the earth's orbit around the sun. It is believed that these variations affect the distribution of solar radiation over the earth and thus cause weather cycles, such as those which have occurred during the Pleistocene epoch.

There are three ways in which the earth's orbit around the sun changes. First, the obliquity of the ecliptic, that is, the angle that the plane through the equator makes with the plane of the earth's orbit around the sun, varies cyclically over an average range of 1.5 degrees about a mean of 23.1 degrees and with an approximate period of 40,000 years. When the obliquity is large, the contrast between seasons is great—winters are cold, and summers are hot; when the obliquity is small, both summers and winters are mild. Changes are in phase in the two hemispheres.

Second, the eccentricity of the earth's orbit fluctuates from 0.00 to 0.05 with a variable period averaging about 97,000 years. Because of this factor, the net solar radiation received by the earth may vary by 20 percent between seasons; however, the integrated annual insolation decreases only slightly more than 1 percent from minimum to maximum eccentricity (Carpenter, 1955). The hemisphere that is closest to the sun in winter has mild winters and summers; that which is closest in summer has cold winters and hot summers. These conditions are most pronounced, and the difference between hemispheres is greatest when the eccentricity is at a maximum. Currently the earth is closest to the sun during the northern hemisphere winter.

Third, because of variations in the orientation of the equatorial plane and the attraction of other planets upon the earth, the time when the earth is closest to the sun advances by about 25 min each year. Thus, the northern hemisphere, which, because of the eccentricity of the earth's orbit, should now have mild seasons, will be subjected to hot summers and cold winters in 10,500 years, when the earth is closest to the sun in July.

Milankovitch (1941) and Brouwer and Van Woerkom (1953) compute the orbital changes and relate them to climatic variations, assuming that changes of mean temperature are based on simple radiative equilibrium with respect to changing insolation and that glaciation is favored by mild seasons, that is, by small obliquity with high eccentricity and the earth closest to the sun in winter. Extending his record back 600,000 years, at 5,000-year intervals, and neglecting effects of the atmospheric circulation, Milankovitch obtains temperature departures from the present normals as large as 7°C at higher latitudes at the times of the solstices. Negative summer departures are balanced, however, by positive winter departures, giving no net change of the annual insolation for either hemisphere. Milankovitch finds that temperature effects of the orbital changes are opposite poleward and equatorward of 43°. During the ice ages, temperatures are above normal (relative to the present) in winter, especially in low latitudes, and below normal in summer, especially in high latitudes. The poleward temperature gradient is thus intensified by above normal annual temperatures equatorward of 43° and below normal annual temperatures poleward of 43°. During interglacial ages, the polar regions are relatively warm and the tropics relatively cool.

There have been many objections to the astronomical explanation for the ice ages. Nevertheless, it has been accepted by several scientists, among them Panofsky (1956), Emiliani (1958), Zeuner (1959, 1961), Jardetzky (1961), Karlstrom (1961), and Fairbridge (1961). Panofsky prefers the theory for the novel reason that orbital changes must have occurred, whereas long-period variations of solar radiation are entirely speculative.

Among the objections, the following are the ones that are most commonly quoted.

1. The glacial-interglacial cycle is extended back indefinitely in time. This implies that some other factor, such as volcanism, mountain building, or variations in the intensity of solar radiation, must first act to create conditions favorable for a glacial epoch before orbital changes can have any effect in producing ice ages.
2. The climatic optimum is dated at 10,000 years ago, whereas it actually occurred about 6,000 years ago. This error is too large to be accidental, unless, as has been recently proposed, it took about 5,000 years for the increased insolation to warm the oceans and atmosphere to a maximum (Fairbridge, 1961). Reasoning of this sort, although physically sound, is a case of attempting to modify the basic theory to fit the facts. Most statisticians would frown on such a practice and point out that almost any variable could be forced to agree moderately well with the essentially random climatic oscillations of the past 600,000 years.
3. According to one interpretation, the length of both the Pleistocene epoch and its ice ages are made much shorter than geological evidence seems to indicate. Unfortunately, current dating techniques are too inaccurate for us to say that the astronomical theory is completely wrong on this point. Emiliani (1958) accepts it and, on the basis of paleo-temperature measurements from six deep-sea cores taken in the Atlantic, Caribbean, and Mediterranean, has constructed a composite temperature curve for the past half-million years, placing the first ice age at 275,000 years B.P. (Table 28).
4. Possible climatic cycles with periods of less than 21,000 years cannot be explained. This objection does not make the astronomical theory invalid, but, together with the first objection, implies that this theory must be used mainly to account for the ice ages and the major readvances. Its failure to explain cycles of 21,000 years or less is not critical, especially since the existence of such cycles is somewhat dubious.
5. Contrary to the evidence that all of the major glaciations have been worldwide, variations in the earth's orbit would produce opposite climatic changes in the two hemispheres. As pointed out by Panofsky (1956), these changes would be out of phase by only 10,000 years, which geologically speaking is a relatively insignificant time period and one which is very difficult to resolve by modern dating techniques. In addition, Fairbridge (1961, 1963) notes that 95 percent of the world's mountain glaciers lie in middle latitudes of the north-

ern hemisphere; therefore, the hydrologic balance of the earth and especially variations in sea level are governed by climatic events in these "sensitive" latitudes.

6. In spite of Milankovitch's results, which indicate that orbital changes could cause temperature departures from the present normals of as large as 7°C, there is still considerable doubt about the magnitude of the effects. Simpson (1934) reworked Milankovitch's calculations, assuming selective rather than gray-body transmission of solar radiation by the earth's atmosphere, and obtained temperature changes that would be too small to have any noticeable influence on the climate. Simpson has been criticized in turn for minimizing the effect of heat storage in the oceans. Perhaps this point can best be resolved by noting that the northern hemisphere should now be having mild winters and summers, since the earth is closest to the sun in January. Hence, it appears that astronomical effects have been overshadowed by the influence of continentality which has produced, by and large, a severe climate, with cold winters and hot summers, over most of the land areas of the northern hemisphere.

Conclusions.—At the present time, a critical evaluation of the many theories of climatic change that have been presented is hampered by our lack of knowledge of the physical environment in which we live. For this reason, I prefer only to state briefly which of the theories discussed in this chapter seem to me to have played major roles in determining the past climate of the earth. Actually, there is a place for almost all of them in the sequence of climatic change. Only the variable sun theories of Öpik and Simpson and the Ewing-Donn theory are considered too speculative to be acceptable at this time. Of the first two, the Öpik theory of glaciation with a cool sun is most appealing, but only as a possible mechanism for setting the stage for the glacial epochs.

The anomalous warmth north of 40°N during the Paleozoic and Mesozoic eras is explained quite satisfactorily by continental drift, the evidence for which is almost indisputable. The cause or causes of the glacial epochs are not so obvious; however, the location of considerable land masses near the poles, mountain building, and the eustatic lowering of sea level seem to be necessary, the long time delay between orogenesis and glaciation perhaps being aided by an initial excess of atmospheric carbon dioxide, as suggested by Plass. The retreat and almost complete disappearance of the ice sheets periodically within a glacial epoch may also be due to an accumulation of atmospheric carbon dioxide, but, most likely, variations in the earth's orbit around the sun are more important.

Climatic changes with periods shorter than from 10,000 to 20,000 years may be caused either by variations in sunspot or other solar activity, tidal-force intensity, atmospheric carbon dioxide, or perhaps volcanism. The next few decades will be a good test of the importance of Willett's sunspot theory, since the period from 1960 to 1980 should be one of reduced sunspot activity and, presumably, glacial advance.

According to the carbon dioxide theory, on the other hand, the warmth of the first half of the present century should continue.

Before much more progress can be made in explaining the causes of climatic variations, we must know more about the earth's interior and the forces that have caused the continents to drift. We must also know more about the general circulation of the atmosphere and how it is affected by the underlying topography, by variations in the solar intensity, and by variations in the spectral distribution of radiation from the sun. Finally, we must know more about the composition of the sun and the factors that have caused its radiative output to vary.

Data from satellite observations will be of considerable value in answering many of the questions that now exist, as will numerical solutions of the energy balance equations and of the equations of motion for the atmosphere. Laboratory experiments with scale models of the earth, the earth's atmosphere, and the sun will also be useful. As the answers emerge, it will be important to keep an open mind and to try to integrate the information into a coherent and understandable picture of climatic change as it really happened.

APPENDIXES

1/ ENERGY EQUIVALENTS

One 15°C g-cal cm^{-2} min^{-1} = 1 cal cm^{-2} min^{-1}

= 1 langley (ly) min^{-1}

= 10^{-3} kcal cm^{-2} min^{-1} = 10^{-3} kly min^{-1}

= 10^{3} mcal cm^{-2} min^{-1} = 10^{3} mly min^{-1}

= 4.1855 absolute joules cm^{-2} min^{-1}

= 4.1855×10^{7} ergs cm^{-2} min^{-1}

= 0.99968 international steam table (IT) cal cm^{-2} min^{-1}

= 69.75 milliwatt (mw) cm^{-2}

= 3.6855 British thermal units (BTU) ft^{-2} min^{-1}

2/ DAILY TOTAL SOLAR RADIATION INCIDENT ON A HORIZONTAL SURFACE AT THE TOP OF THE ATMOSPHERE

From equations (3.1) and (3.2),

$$Q_s = \int_{-H}^{H} Q'_s dt = S\left(\frac{\bar{d}}{d}\right)^2 \int_{-H}^{H} (\sin\varphi \sin\delta + \cos\varphi \cos\delta \cos h)\, dt.$$

But $dh/dt = \omega$ = angular velocity of the earth = 2π rad day^{-1}. Hence,

$$Q_s = S\left(\frac{\bar{d}}{d}\right)^2 \int_{-H}^{H} (\sin\varphi \sin\delta + \cos\varphi \cos\delta \cos h)\, \frac{dh}{\omega}$$

or

$$Q_s = \frac{1{,}440}{\pi} S\left(\frac{\bar{d}}{d}\right)^2 (H \sin\varphi \sin\delta + \cos\varphi \cos\delta \sin H) \text{ ly day}^{-1}$$

where the solar constant is in langleys per minute.

An alternative form may be obtained by multiplying both sides of equation (3.3) by $\tan H$ and showing that

$$\cos\varphi \cos\delta \sin H = -\sin\varphi \sin\delta \tan H.$$

Then

$$Q_s = \frac{1{,}440}{\pi} S\left(\frac{\bar{d}}{a}\right)^2 (H - \tan H) \sin\varphi \sin\delta.$$

3/ EFFECT OF AIR-GROUND TEMPERATURE DIFFERENCES ON THE EFFECTIVE OUTGOING RADIATION

Let I be the effective outgoing radiation from the surface, as estimated from radiation charts or equations with temperature and humidity data taken at screen height; let I_s be the true effective outgoing radiation from the surface; and let I' be the true net radiative flux at screen height. It then follows that

$$I = I_{\uparrow} - I_{\downarrow} = \epsilon\sigma T^4 - I_{\downarrow} \tag{3a}$$

$$I_s = I_{\uparrow s} - I_{\downarrow s} = \epsilon\sigma T_s^4 - I_{\downarrow s} \tag{3b}$$

and

$$I' = (1 - \epsilon_L)\sigma T_s^4 + \epsilon_L\sigma T_L^4 - I_{\downarrow}/\epsilon \tag{3c}$$

where variables without subscripts refer to screen height; variables with the subscript s refer to the surface; and variables with the subscript L refer to the air layer below screen height. In equation (3c), the net radiative flux through the screen height level is expressed as the sum of three terms, the first representing the energy transmitted through the level directly from the surface, the second the upward flux of energy emitted by the layer itself, and the third the atmospheric counter radiation through the level.

Combining equations (3a) and (3b), we get

$$I_s - I = \epsilon\sigma(T_s^4 - T^4) - (I_{\downarrow s} - I_{\downarrow}) . \tag{3d}$$

But

$$I_s - \epsilon I' = \epsilon\epsilon_L\sigma(T_s^4 - T_L^4) - (I_{\downarrow s} - I_{\downarrow})$$

and

$$I_s - I = \epsilon\sigma(T_s^4 - T^4) - \epsilon\epsilon_L\sigma(T_s^4 - T_L^4) + I_s - \epsilon I'$$

or

$$I_s - I = \epsilon\sigma(T_s^4 - T^4)\left[1 + \frac{I_s/\epsilon - I'}{\sigma(T_s^4 - T^4)} - \frac{\epsilon_L(T_s^4 - T_L^4)}{(T_s^4 - T^4)}\right]. \qquad (3e)$$

The magnitude of the second term in the brackets may be estimated from observations made by Rider and Robinson (1951) at the Kew Observatory in England. Their data show that when $\epsilon = 1.0$, this term varies from 0.006 to 0.205 and averages 0.079 ± 0.037, or approximately 0.08, when screen height is at 1 m.

The temperature ratio in the last term in the brackets may be closely approximated by $(T_s - T_L)/(T_s - T)$ and under normal conditions close to the ground, where the temperature varies nearly logarithmically with height (see chap. 10), will equal about 0.8 for relatively smooth surfaces. That is, the mean temperature of the layer is much closer to the temperature at screen height than it is to the surface temperature.

Large temperature differences between the surface and 1 m above the surface will normally occur only when the air is very dry, that is, when the midday relative humidity is less than 20 percent. Under these conditions, the layer would certainly contain no more than 2×10^{-3} cm of water vapor and 3×10^{-2} cm of carbon dioxide. The two gases, when combined, would then give a layer emissivity ϵ_L of about 17 percent (see Table 8*a*).

Hence, the last term in the brackets in equation (3e) is about 0.14, tending to cancel the second term. Therefore,

$$I_s - I \simeq \epsilon\sigma(T_s^4 - T^4) \simeq 4\epsilon\sigma T^3(T_s - T) .$$

The latter approximation is very good. Comparing this result with equation (3d), we see that neglecting the last two terms in equation (3e) is equivalent to assuming that the counter radiation does not change between screen height and the ground.

The quantity $I_s/\epsilon - I'$, which appears in equation (3e), represents the net radiative flux into the air layer between screen height and the ground when the surface emissivity is 1.0. As such, it is directly related to the radiative heating or cooling of the layer, through equation (3.11). From the data of Rider and Robinson (1951)

$$I_s - I' \simeq 0.08\sigma(T_s^4 - T^4) \simeq 0.32\sigma T^3(T_s - T)$$

and from equation (3.11)

$$\frac{\Delta \bar{T}_I}{\Delta t} \simeq \frac{0.32\sigma T^3(T_s - T)}{\rho c_p \Delta z}$$

where c_p is the specific heat of air at constant pressure.

4/ GENERAL PRINCIPLES OF PYRANOMETER AND RADIOMETER OPERATION

PYRANOMETERS

Consider two sensing elements, one black and one white, under a glass hemisphere. Assume, for simplicity, that the black element absorbs all incident radiation, both shortwave $(Q + q)$ and longwave (σT_g^4, where T_g is the temperature of the glass whose infrared emissivity is assumed to equal 1.0), and that the white element absorbs only the longwave radiation (σT_g^4). In both cases, the absorbed energy will either be stored in the sensing element, thus raising its temperature, or be radiated outward. There will also be some heat conduction from the elements to the air; but because the sensors are protected from the natural turbulence and convection of the free air by the glass dome, this is normally small compared with the other components and will be neglected. If the storage rate is G, the energy balance equation for each element is

$$\text{(black element)}\quad (Q + q) + \sigma T_g^4 = G_b + \sigma T_b^4$$

$$\text{(white element)}\quad \sigma T_g^4 = G_w + \sigma T_w^4 .$$

Subtracting gives

$$Q + q = G_b - G_w + \sigma(T_b^4 - T_w^4) . \tag{4a}$$

The storage rates G_b and G_w are proportional to the temperature differences, $T_b - T_r$ and $T_w - T_r$, respectively, between the upper and lower faces of the sensing elements. The constant of proportionality for each element is its thermal conductivity λ divided by its thickness d. If this constant is the same for both elements and their lower faces are at the same temperature, it follows that

$$G_b - G_w = \frac{\lambda}{d}\,(T_b - T_w) . \tag{4b}$$

Approximating $(T_b^4 - T_w^4)$ by $4T_w^3(T_b - T_w)$, we see that equation (4*a*) then becomes

$$Q+q=\left(\frac{\lambda}{d}+4\sigma T_w^3\right)(T_b-T_w)$$

or

$$Q+q=K(T_b-T_w)$$

where the constant of proportionality K is a function of the physical properties and temperature of the sensor. The temperature dependence is weak, however, and the same constant may be used over the entire range of temperatures likely to be encountered without introducing serious errors.

NET RADIOMETERS

Consider two freely exposed, blackened sensing elements placed back to back, with one facing upward and the other facing downward. Assume that there is a thin layer of insulating material between the two elements. Let T_u, T_d, and T_m be the temperature of the exposed face of the upper element, the temperature of the exposed face of the lower element, and the temperature of the insulating material.

Both elements are assumed to absorb all radiation incident upon them, which for the upper element is the total solar radiation $Q + q$ plus the infrared counter radiation $I_\downarrow$ and for the lower element is the solar radiation reflected from the underlying surface $(Q + q)a$ plus the longwave terrestrial radiation $I_\uparrow$. The absorbed energy is either stored in the sensing elements, thus raising their temperatures; radiated outward; or lost to the free air by conduction and convection. If the storage rate is again represented by G and the rate of convective heat loss by H, the energy balance equation for each element is

$$\text{(upper element) } Q+q+I_\downarrow = G_u + H_u + \sigma T_u^4 \tag{4c}$$

$$\text{(lower element) } (Q+q)a+I_\uparrow = G_d + H_d + \sigma T_d^4 \tag{4d}$$

Subtraction gives

$$(Q+q)(1-a) - I_\uparrow + I_\downarrow = G_u - G_d + H_u - H_d + \sigma(T_u^4 - T_d^4)\,. \tag{4e}$$

The storage rates G_u and G_d are proportional to the temperature differences, $T_u - T_m$ and $T_d - T_m$, respectively, between the outer and inner surfaces of the elements. If the constant of proportionality λ/d is the same for both elements, it follows, as in equation (4*b*), that

$$G_u - G_d = \frac{\lambda}{d}\,(T_u - T_d)\,. \tag{4f}$$

The convective heat loss rates H_u and H_d are proportional to the temperature differences $T_u - T$ and $T_d - T$, respectively, between the exposed surfaces of the elements and the free air, assumed here to be at a constant temperature T. The constant of proportionality $f(u)$ is a function primarily of the free-air wind speed, which is assumed to be the same across both elements. Hence

$$H_u - H_d = f(u)(T_u - T_d)\,. \tag{4g}$$

Recognizing the left-hand side of equation (4*e*) as the radiation balance R; replacing $(T_u^4 - T_d^4)$ by $4T_d^3(T_u - T_d)$; and using equations (4*f*) and (4*g*), we get

$$R = \left[\frac{\lambda}{d} + f(u) + 4\sigma T_d^3\right](T_u - T_d)\,. \tag{4h}$$

If the lower element is covered with highly reflective polished aluminum, so that its absorptivity and emissivity are negligible at all wavelengths, equation (4*d*) is replaced by

$$\text{(lower element)}\ 0 = G_d + H_d$$

and equation (4*h*) becomes

$$Q + q + I_\downarrow = \left[\frac{\lambda}{d} + f(u)\right](T_u - T_d) + \sigma T_u^4$$

or

$$Q + q + I_\downarrow = \left[\frac{\lambda}{d} + f(u) + 4\sigma T_d^3\right](T_u - T_d) + \sigma T_d^4\,.$$

5/ THE MERIDIONAL FLUX OF WATER VAPOR IN THE ATMOSPHERE

From equation (7.13), we have

$$[\bar{c}_m]A = L[\overline{uw_a}]$$

where the bar represents a time average at a given station on a latitude circle and the brackets represent a latitudinal average. The instantaneous air speed u can be expressed in terms of its deviation u' from $\bar{u}$. Thus

$$u = \bar{u} + u' .$$

Similarly,

$$w_a = \bar{w}_a + w_a' .$$

Since the average values of both w_a' and u' are zero, it follows that

$$\bar{c}_m \frac{A}{L} = \overline{uw_a} = \bar{u}\bar{w}_a + \overline{u'w'_a} . \tag{5a}$$

The first term on the right of equation (5*a*) represents the part of the total vapor flux associated with a mean north-south air motion at the station. The second term gives the flux associated with the day-to-day fluctuations of w_a and u. This second term will be zero unless south winds carry more moisture over the station than do north winds, or vice versa.

Averaging equation (5*a*) over the latitude circle and representing the averaging process by brackets, we get

$$[\bar{c}_m] \frac{A}{L} = [\bar{u}\bar{w}_a] + [\overline{u'w_a'}]. \tag{5b}$$

If $\bar{u} = [\bar{u}] + \bar{u}^*$ and $\bar{w}_a = [\bar{w}_a] + \bar{w}_a^*$, where the asterisked quantities represent deviations from the latitudinal mean,

$$[\bar{u}\bar{w}_a] = [\bar{u}][\bar{w}_a] + [\bar{u}^*\bar{w}_a^*]$$

since the latitudinal averages of $\bar{w}_a^*$ and $\bar{u}^*$ are zero. Hence, equation (5*b*) becomes

$$[\bar{c}_m]\frac{A}{L} = [\bar{u}][\bar{w}_a] + [\bar{u}^*\bar{w}_a^*] + [\overline{u'w_a'}]$$

or

$$[\bar{c}_m] = [\bar{c}_m]_1 + [\bar{c}_m]_2 + [\bar{c}_m]_3 \; .$$

6/ CALCULATION OF MONTHLY AVERAGE VALUES OF THE SURFACE ENERGY BALANCE COMPONENTS FROM CLIMATOLOGICAL DATA

LAND STATIONS

The net radiation was determined from the formula

$$R = R_h - 4\epsilon\sigma T^3(T_s - T)$$

where

$$R_h = (Q + q)(1 - a) - I_o(1 - kn^2)\,.$$

The total incoming solar radiation $Q + q$ was obtained either from pyranometer observations or from maps, such as those shown in Figures 9 and 10. The surface albedo a was estimated for the ground cover typical of the area. We can read the effective outgoing radiation with clear skies I_o from Figure 16, using air temperatures from published climatological data and precipitable water vapor values from Reitan (1960*a*, *b*). The surface emissivity was assumed to be 0.9.

The parameter k was obtained from the accompanying table in which

$$v = 1 - \frac{1}{n}\left[1 - \frac{Q+q}{(Q+q)_o}\right] = \frac{(Q+q)_c}{(Q+q)_o}$$

where $(Q + q)_o$ is the total solar radiation with clear skies and $(Q + q)_c$ is the total

v	k	v	k
0.1	1.000	0.6	0.490
.2	0.985	.7	.320
.3	0.920	.8	.195
.4	0.815	0.9	0.115
0.5	0.660		

solar radiation with overcast skies ($n = 1.0$). The quantities v and k are related through their joint dependence on the cloud height, as given for k in chapter 4 and for v by Haurwitz (1948). The term $(Q + q)_o$ was determined from graphs similar to that shown in Figure 11 for Inyokern, California. This method was suggested by Budyko (1956).

We can estimate the term correcting from the temperature difference $T_s - T$ between the surface and the air at 1 or 2 m by assuming that

$$H = b(T_s - T)$$

and writing the energy balance equation in the form

$$4\epsilon\sigma T^3(T_s - T) = \frac{R_h - G - LE}{1 + \dfrac{b}{4\epsilon\sigma T^3}}.$$

According to Budyko (1956), the ratio $b/4\epsilon\sigma T^3$ has a mean value of about three when the temperature decreases with height (lapse conditions) or, equivalently, when $R_h - G - LE > 0$. With inversions, when the temperature increases with height and $R_h - G - LE < 0$, the ratio averages about one.

The heat flux into the soil G was estimated from

$$G(0, t) - \Delta T_0(\omega C\lambda)^{1/2} \sin\left(\omega t + \frac{\pi}{4}\right),$$

which is equation (9.16). Appropriate values of the thermal property $(C\lambda)^{1/2}$ were used, and it was assumed that $t = 0$ on April 15. The energy used for evaporation was determined by the water balance method described in chapter 11, pages 175–76. The sensible heat flux was taken as the residual in the energy balance equation.

OCEAN STATIONS

The data for R, LE, and H are taken from Budyko (1963). He obtains the net radiation from climatological data and the formula

$$R = (Q + q)_o(1 - \alpha)(1 - an - bn^2) - \epsilon\sigma T_s^4(11.7 - 0.23e_s)(1 - cn)$$

where e_s is the saturation vapor pressure in millibars at the surface temperature T_s; b is a constant and equals 0.38; and a and c are parameters that depend on the cloud height. For a and c, Budyko uses the average latitudinal values listed on page 242. He also uses average latitudinal values of $(Q + q)_o$ and α for each month. These are in his publication.

The latent and sensible heat fluxes into the atmosphere were determined from ship observations and the relationships

$$E = 0.134u(e_s - e) \text{ mm day}^{-1}$$

$$L = 597 - 0.56T_s \text{ cal g}^{-1}$$

$$H = 5.18u\,(T_s - T) \text{ ly day}^{-1}\,.$$

Here u and T are, respectively, the deck-level wind speed in meters per second and temperature in degrees Centigrade. These formulas are similar to those used by Jacobs (1951), although the coefficients are about 5 percent smaller.

Latitude	*a*	*c*	Latitude	*a*	*c*
0°	0.38	....	45°	0.38	....
5°	.40	0.50	50°	.40	0.72
10°	.40	0.52	55°	.41	....
15°	.39	0.55	60°	.36	0.76
20°	.37	0.59	65°	.25	....
25°	.35	....	70°	.18	0.80
30°	.36	0.63	75°	.16	....
35°	.38	....	80°	.15	0.84
40°	0.38	0.68	85°	0.14	0.86

Heat storage in the oceans, G, was approximated by a method suggested by Gabites (1950). Assuming that the amplitude of the annual temperature change is constant and equal to the surface value in the upper 25 m of ocean and decreases linearly to zero at a depth of 125 m, he writes, using equation (3.11)

$$G = 7{,}500 \frac{\Delta T_s}{\Delta t} \text{ ly day}^{-1}$$

where $\Delta T_s/\Delta T$ is the rate of change of surface temperature in °C day^{-1}. He lists mean monthly and latitudinal values of $\Delta T_s/\Delta T$. These were adjusted for the annual amplitude of T_s at the locations in Figure 32.

The net transfer of heat by ocean currents was obtained as the residual in the energy balance equation.

REFERENCES

Abraham, F. F. 1960. Determination of Long-Wave Atmospheric Radiation. *J. Meteorol.* 17:291–95.

Abramowitz, M., and Stegun, I. A. 1964. *Handbook of Mathematical Functions with Formulas, Graphs, and Mathematical Tables.* Applied Math Series No. 55. U.S. Government Printing Office, Washington.

Albrecht, F. 1961. *Der jährliche Gang der Komponenten des Wärme- und Wasserhaushaltes der Ozean.* Berichte Deutschen Wetterdienstes No. 79. Bad Kissingen.

Albrecht, F. 1962. *Die Berechnung der natürlichen Verdunstung (Evapotranspiration) der Erdoberfläche aus klimatologischen Daten.* Berichte Deutschen Wetterdienstes No. 83. Bad Kissingen.

Ali, B. 1932. Variation of Wind with Height. *Quart. J. Roy. Meteorol. Soc.* 58:285–88.

Anderson, E. R. 1954. Energy-Budget Studies, pp. 71–119. In *Water-Loss Investigations: Lake Hefner Studies, Technical Report.* U.S. Geological Survey Professional Paper 269, Washington.

Ångström, A. 1916. Über die Gegenstrahlung der Atmosphäre. *Meteorol. Z.* 33: 529–38.

Antevs, E. 1955. Geologic-Climatic Method of Dating, pp. 151–69. In *Geochronology.* Physical Sciences Bull. No. 2, The University of Arizona, Tucson.

Arnold, C. A. 1947. *An Introduction to Paleobotany.* McGraw-Hill Book Co., Inc., New York.

Aslyng, H. C. 1960. Evaporation and Radiation Heat Balance at the Soil Surface. *Arch. Meteorol., Geophys. Bioklimatol., Ser. B.* 10:359–75.

Astling, E. G., and Horn, L. H. 1964. Some Geographical Variations of Terrestrial Radiation Measured by TIROS II. *J. Atmos. Sci.* 21:30–34.

Bannon, J. K., and Steele, L. P. 1960. *Average Water-Vapour Content of the Air.* Geophysical Mem. No. 102. Meteorol. Office, Air Ministry, London.

Barad, M. L. 1963. Examination of a Wind Profile Proposed by Swinbank. *J. Appl. Meteorol.* 2:747–54.

Barad, M. L., and Fuquay, J. J. 1962. *The Green Glow Diffusion Program*. Vols. I–II. Geophysics Research Papers No. 73. Air Force Cambridge Research Center, Bedford.

Barad, M. L., and Haugen, D. A. (eds.) 1958. *Project Prairie Grass, A Field Program in Diffusion*. Vols. I–III. Geophysics Research Papers No. 59. Air Force Cambridge Research Center, Bedford.

Barad, M. L., and Haugen, D. A. 1959. A Preliminary Evaluation of Sutton's Hypothesis for Diffusion from a Continuous Point Source. *J. Meteorol.* 16:12–20.

Baumgartner, A. 1953. Das Eindringen des Lichtes in den Boden, *Forstwiss. Zentr.* 72:172–84.

Baumgartner, A. 1956. *Untersuchungen über den Wärme- und Wasserhaushalt eines jungen Waldes*. Berichte Deutschen Wetterdienstes No. 28. Bad Kissingen.

Baur, F., and Philipps, H. 1935. Der Wärmehaushalt der Lufthülle der Nordhalbkugel im Januar und Juli und zur Zeit der Äquinoktien und Solstitien. Pt. II. Ausstrahlung, Gegenstrahlung und meridionaler Wärmetransport bei normaler Solarkonstante. *Beitr. Geophys.* 45:82–132.

Bavel, C. H. M. van, and Fritschen, L. J. 1964. *Energy Balance Studies over Sudan Grass, 1962*. Interim Report. U.S. Water Conservation Laboratory, Tempe.

Bavel, C. H. M. van, Fritschen, L. J., and Reeves, W. E. 1963. Transpiration by Sudan Grass as an Externally Controlled Process. *Science* 141:269–70.

Bavel, C. H. M. van, and Myers, L. E. 1962. Automatic Weighing Lysimeters. *J. Am. Soc. Agr. Engr.* 43:580–83, 586–88.

Bell, B. 1953. Solar Variation as an Explanation of Climate Change, pp. 123–26. In *Climatic Change*. Harvard University Press, Cambridge.

Benton, G. S., Blackburn, R. T., and Snead, V. O. 1950. The Role of the Atmosphere in the Hydrologic Cycle. *Trans., Am. Geophys. Union* 31:61–73.

Benton, G. S., and Estoque, M. A. 1954. Water-Vapor Transfer over the North American Continent. *J. Meteorol.* 11:462–77.

Berlyand, T. G. 1956. Teplovoi Balans Atmosfery Severnogo Polushariya. In *A. I. Voeikov i Sovremennye Problemy Klimatologii*. Gidrometeorologicheskoe Izdatel'stvo, Leningrad. (English transl.: IPST Staff 1963. Heat Balance of the Atmosphere of the Northern Hemisphere, pp. 57–83. In *A. I. Voeikov and Problems in Climatology*. Office of Technical Services, U.S. Dept. of Commerce, Washington.)

Best, A. C. 1935. *Transfer of Heat and Momentum in the Lowest Layers of the Atmosphere*. Geophysical Mem. No. 65. Meteorol. Office, Air Ministry, London.

Bierly, E. W., and Hewson, E. W. 1962. Some Restrictive Meteorological Conditions To Be Considered in the Design of Stacks. *J. Appl. Meteorol.* 1:383–90.

Bigg, E. K., and Miles, G. T. 1964. The Results of Large-Scale Measurements of Natural Ice Nuclei. *J. Atmos. Sci.* 21:396–403.

Bijl, W., van der. 1958. *The Evapotranspiration Problem, First Contribution*. Report 1956–1957. Kansas State College Agri. Exptl. Station, Manhattan.

Blackwell, J. H. 1954. A Transient-Flow Method for Determination of Thermal Constants of Insulating Materials in Bulk. *J. Appl. Phys.* 25:137–44.

Blaney, H. F. 1955. Evaporation Study at Silver Lake in Mojave Desert, California. *Trans. Am. Geophys. Union* 38:209–15.

Blaney, H. F., and Criddle, W. D. 1962. *Determining Consumptive Use of Irrigation Water Requirements*. Tech. Bull. 1275. U.S. Dept. of Agriculture, Washington.

Brier, G. W. 1961. A Test of the Reality of Rainfall Singularities. *J. Meteorol.* 18: 242–46.

Brier, G. W., and Bradley, D. A. 1964. The Lunar Synodical Period and Precipitation in the United States. *J. Atmos. Sci.* 21:386–95.

British Meteorological Office. 1956. *Handbook of Meteorological Instruments. Pt. I. Instruments for Surface Observations*. Her Majesty's Stationery Office, London.

Brooks, C. E. P. 1927. The Mean Cloudiness over the Earth. *Mem. Roy. Meteorol. Soc.* 1:127–38.

Brooks, C. E. P. 1951. Geological and Historical Aspects of Climatic Change, pp. 1004–18. In *Compendium of Meteorology*. American Meteorological Society, Boston.

Brooks, C. E. P., and Hunt, T. M. 1930. The Zonal Distribution of Rainfall over the Earth. *Mem. Roy. Meteorol. Soc.* 3:139–58.

Brooks, F. A. 1959. *An Introduction to Physical Micrometeorology*. University of California Press, Davis.

Brunt, D. 1932. Notes on Radiation in the Atmosphere. *Quart. J. Roy. Meteorol. Soc.* 58:389–420.

Brutsaert, W. 1965. Evaluation of Some Practical Methods of Estimating Evapotranspiration in Arid Climates at Low Latitudes. *Water Resources Research* 1: 187–91.

Budyko, M. I. 1956. *Teplovoi Balans Zemnoi Poverkhnosti*. Gidrometeorologicheskoe Izdatel'stvo, Leningrad. (English transl.: Stepanova, N. A. 1958. *The Heat Balance of the Earth's Surface*. Office of Technical Services, U.S. Dept. of Commerce, Washington.)

Budyko, M. I. (ed.) 1963*a*. *Atlas Teplovogo Balansa*. Gidrometeorologicheskoe Izdatel'skoe, Moscow.

Budyko, M. I. 1963*b*. *Evaporation under Natural Conditions*. Office of Technical Services, Washington. (Transl. from Russian.)

Budyko, M. I., and Kondratiev, K. Y. 1964. The Heat Balance of the Earth, pp. 529–54. In *Research in Geophysics. Vol. II. Solid Earth and Interface Phenomena*. M.I.T. Press, Cambridge.

Budyko, M. I., Yefimova, N. A., Zubenok, L. I., and Strokina, L. A. 1962. The Heat Balance of the Surface of the Earth. *Sov. Geograph.* 3:3–16.

Buettner, K. 1955. A Small Portable Meter for Soil Heat Conductivity and Its Use in the O'Neill Test. *Trans. Am. Geophys. Union* 36:827–30.

Buettner, K. J. K., and Kern, C. D. 1963. Infrared Emissivity of the Sahara from Tiros Data. *Science* 142:671.

Burgos, J. J., and Tschapek, M. 1958. Water Storage in Semi-Arid Soils, pp. 72–92. In *Climatology and Microclimatology*. UNESCO, Paris.

Businger, J. A. 1959. A Generalization of the Mixing-Length Concept. *J. Meteorol.* 16:516–23.

Businger, J. A., and Buettner, K. J. K. 1961. Thermal Contact Coefficient (A Term Proposed for Use in Heat Transfer). *J. Meteorol.* 18:422.

Butzer, K. W. 1958. *Quaternary Stratigraphy and Climate in the Near East*. Bonner Geographische Abhandlungen, Heft 24. Ferd. Dümmlers Verlag, Bonn.

Callendar, G. S. 1958. On the Amount of Carbon Dioxide in the Atmosphere. *Tellus* 10:243–48.

Carpenter, E. F. 1955. Astronomical Aspects of Geochronology, pp. 29–74. In *Geochronology*. Physical Sciences Bull. No. 2, The University of Arizona, Tucson.

Carslaw, H. S., and Jaeger, J. C. 1959. *Conduction of Heat in Solids*. Clarendon Press, Oxford.

Carson, J. E. 1961. *Soil Temperature and Weather Conditions*. Rep. No. 6470, Argonne National Laboratories, Argonne.

Cartwright, G. D., and Rubin, M. J. 1961. Inside Antarctica No. 6-Meteorology at Mirny. *Weatherwise* 14:110–18.

Castany, G. 1963. *Traité Pratique des eaux Souterraines*. Dunod, Paris.

Chang, J. 1958. *Ground Temperature*. Vol. I. Harvard University, Blue Hill Meteorological Observatory, Milton.

Clayton, H. H., and Clayton, F. L. (eds.) 1944, 1947. *World Weather Records*. Smithsonian Miscellaneous Collections, Vols. 79, 90, 105. Smithsonian Institution, Washington.

Conover, J. H. (ed.) 1959. *World Weather Records 1941–50*. U.S. Government Printing Office, Washington.

Courvoisier, P., and Wierzejewski, H. 1954. Das Kugelpyranometer Bellani. *Arch. Meteorol. Geophys., Bioklimatol. Series B*, 5:413–46.

Covey, W., Halstead, M. H., Hillman, S., Merryman, J. D., Richman, R. L., and York, A. H. 1958. Micrometeorological Data Collected by Texas A & M, pp. 53–96. In *Project Prairie Grass, A Field Program in Diffusion*. Vol. II. Geophysical Research Papers No. 59. Air Force Cambridge Research Center, Bedford.

Craig, R. A. 1952. Surface Pressure Variations Following Geomagnetically Disturbed and Geomagnetically Quiet Days. *J. Meteorol.* 9:126–38.

Cramer, H. E., 1959*a*. Engineering Estimates of Atmospheric Dispersal Capacity. *Am. Ind. Hyg. Assoc. J.* 20:183–89.

Cramer, H. E. 1959*b*. A Brief Survey of the Meteorological Aspects of Atmospheric Pollution. *Bull. Am. Meteorol. Soc.* 40:165–71.

Crawford, T. V., and Dyer, A. J. 1962. The Vertical Divergence of Evaporative and Sensible Heat Fluxes, pp. 53–75. In *Investigation of Energy and Mass Transfers near the Ground Including the Influences of the Soil-Plant-Atmosphere System.* Second Annual Report. University of California, Davis.

Davidson, B., and Barad, M. L. 1956. Some Comments on the Deacon Wind Profile. *Trans. Am. Geophys. Union* 37:168–76.

Davis, P. A. 1963. An Analysis of the Atmospheric Heat Budget. *J. Atmos. Sci.* 20: 5–22.

Deacon, E. L. 1949. Vertical Diffusion in the Lowest Layers of the Atmosphere. *Quart. J. Roy. Meteorol. Soc.* 75:89–103.

Deacon, E. L. 1953. *Vertical Profiles of Mean Wind in the Surface Layers of the Atmosphere.* Geophysical Mem. No. 91. Meteorol. Office, Air Ministry, London.

Denmead, O. T., and Shaw, R. H. 1962. Availability of Soil Water to Plants as Affected by Soil Moisture Content and Meteorological Conditions. *Agron. J.* 54:385–90.

Doell, R. R., and Cox, A. 1961. Paleomagnetism. *Advan. Geophys.* 8:221–313.

Dorf, E. 1960. Climatic Changes of the Past and Present. *Am. Sci.* 48:341–64.

Dorsey, N. E. 1940. *Properties of Ordinary Water-Substance.* Reinhold Publishing Corp., New York.

Drinkwater, W. O., and Janes, B. E. 1957. Relation of Potential Evapotranspiration to Environment and Kind of Plant. *Trans. Am. Geophys. Union* 38:524–28.

Drosdov, O. A. 1956. Balans Vlagi v Atmosfere. In *A. I. Voeikov i Sovremennye Problemy Klimatologii.* Gidrometeorologicheskoe Izdatel'stvo, Leningrad. (English transl.: IPST Staff 1963. The Water Balance in the Atmosphere, pp. 1–18. In *A. I. Voeikov and Problems in Climatology.* Office of Technical Services, U.S. Dept. of Commerce, Washington.)

Drosdov, O. A. 1961. Moisture Circulation and Its Role in Natural Processes. *Sov. Geograph.* 2:12–24.

Dubois, P. 1929. Nächtliche effektive Ausstrahlung. *Beitr. Geophys.* 22:41–99.

Dunbar, C. O. 1960. *Historical Geology.* McGraw-Hill Book Co., Inc., New York.

Dutton, J. A., and Bryson, R. A. 1962. Heat Flux in Lake Mendota. *Limnol. Oceanog.* 7:80–97.

Dyer, A. J. 1961. Measurements of Evaporation and Heat Transfer in the Lower Atmosphere by an Automatic Eddy-Correlation Technique. *Quart. J. Roy. Meteorol. Soc.* 87:401–12.

Dyer, A. J. 1963. The Adjustment of Profiles and Eddy Fluxes. *Quart. J. Roy. Meteorol. Soc.* 89:276–80.

Dyer, A. J., and Pruitt, W. O. 1962. Eddy-Flux Measurements over a Small, Irrigated Area. *J. Appl. Meteorol.* 1:471–73.

Eichhorn, N. D. 1963. *Implications of Rising Carbon Dioxide Content of the Atmosphere*. The Conservation Foundation, New York.

Elliott, W. P. 1960. An Hypothesis for the Diabatic Mixing-Length. *J. Meteorol.* 17: 680–81.

Elliott, W. P. 1964. The Height Variation of Vertical Heat Flux Near the Ground. *Quart. J. Roy. Meteorol. Soc.* 90:260–65.

Ellison, T. H. 1957. Turbulent Transport of Heat and Momentum from an Infinite Rough Plane. *J. Fluid Mech.* 2:456–66.

Elsasser, W. M. 1942. *Heat Transfer by Infrared Radiation in the Atmosphere*. Harvard Meteorological Studies No. 6. Harvard University Printing Office, Cambridge.

Elsasser, W. M., and Culbertson, M. F. 1960. *Atmospheric Radiation Tables*. Meteorological Monograph, Vol. IV, No. 23. American Meteorological Society, Boston.

Emiliani, C. 1957. Glaciation and Their Causes, pp. 36–42. In *Recent Research in Climatology*. Committee on Research in Water Resources, University of California, La Jolla.

Emiliani, C. 1958. Ancient Temperatures. *Sci. Am.* 198(2):54–63.

Ericson, D. B., Broecker, W. S., Kulp, J. L., and Wollin, G. 1956. Late Pleistocene Climates and Deep-Sea Sediments. *Science* 124:385–89.

Ewing, M., and Donn, W. L. 1956. A Theory of Ice Ages. *Science* 123:1061–66.

Fairbridge, R. W. 1961. Convergence of Evidence on Climatic Change and Ice Ages. *Ann. N.Y. Acad. Sci.* 95:542–79.

Fairbridge, R. W. 1963. Mean Sea Level Related to Solar Radiation during the Last 20,000 Years, pp. 229–42. In *Changes of Climate*. UNESCO, Paris.

Fleagle, R. G., and Businger, J. A. 1963. *An Introduction to Atmospheric Physics*. Academic Press, New York.

Flint, R. F. 1957. *Glacial and Pleistocene Geology*. John Wiley & Sons, Inc., New York.

Flint, R. F., and Deevey, E. S. 1951. Radiocarbon Dating of Late-Pleistocene Events. *Am. J. Sci.* 249:257–300.

Flint, R. F., and Gale, W. A. 1958. Stratigraphy and Radiocarbon Dates at Searles Lake, California. *Am. J. Sci.* 256:689–714.

Frankenberger, E. 1960. *Beiträge zum Internationalen Geophysikalischen Jahr 1957/58*. Berichte Deutschen Wetterdienstes No. 73. Bad Kissingen.

Fritz, S. 1951. Solar Radiant Energy and Its Modification by the Earth and Its Atmosphere, pp. 13–33. In *Compendium of Meteorology*. American Meteorological Society, Boston.

Funk, J. P. 1960. Measured Radiative Flux Divergence Near the Ground at Night. *Quart. J. Roy. Meteorol. Soc.* 86:382–89.

Fuquay, J. J., Simpson, C. L., and Hinds, W. T. 1964. Prediction of Environmental

Exposures from Sources Near the Ground Based on Hanford Experimental Data. *J. Appl. Meteorol.* 3:761–70.

Fuquay, J. J., Simpson, C. L., Barad, M. L., and Taylor, J. H. 1963. Results of Recent Field Programs in Atmospheric Diffusion. *J. Appl. Meteorol.* 2:122–28.

Gabites, J. F. 1950. *Seasonal Variations in the Atmospheric Heat Balance.* Sc.D. Dissertation. Massachusetts Institute of Technology, Cambridge.

Gamow, G. 1964. *A Star Called the Sun.* Viking Press, New York.

Gates, D. M. 1962. *Energy Exchange in the Biosphere.* Harper and Row, New York.

Gates, D. M. 1963. The Energy Environment in Which We Live. *Am. Sci.* 51:327–48.

Gates, D. M., Keegan, H. J., Schleter, J. C., and Weidner, V. R. 1965. Spectral Properties of Plants. *Appl. Opt.* 4:11–20.

Geiger, R. 1959. *The Climate Near the Ground.* Harvard University Press, Cambridge. (Transl. by M. N. Stewart.)

Geiger, R. 1961. *Das Klima der bodennahen Luftschicht.* Friedr. Vieweg & Sohn, Braunschweig. (English transl.: Scripta Technica, Inc. 1965. *The Climate near the Ground.* Harvard University Press, Cambridge.)

Gentilli, J. 1961. Quaternary Climates of the Australian Region. *Ann. N.Y. Acad. Sci.* 95:465–501.

Gifford, F., Jr. 1957. Further Data on Relative Atmospheric Diffusion. *J. Meteorol.* 14:475–76.

Goldsmith, J. R. 1962. Effects of Air Pollution on Humans, pp. 335–86. In *Air Pollution.* Vol. I. Academic Press, New York.

Goody, R. M. 1964. *Atmospheric Radiation.* Vol. I. *Theoretical Basis.* The Clarendon Press, Oxford.

Green, R. 1961. Paleomagnetic Significance of Evaporites, pp. 61–88. In *Descriptive Paleoclimatology.* Interscience Publishers, Inc., New York.

Haltiner, G. J., and Martin, F. L. 1957. *Dynamical and Physical Meteorology.* McGraw-Hill Book Co., Inc., New York.

Hanson, K. J. 1960. Radiation Measurements on the Antarctic Snowfield, a Preliminary Report. *J. Geophys. Res.* 65:935–46.

Hapgood, C. H. 1959. The Earth's Shifting Crust, pp. 22, 66–69. In *Saturday Evening Post,* Jan. 10.

Harbeck, G. E. 1958. Results of Energy-Budget and Mass-Transfer Computations, pp. 35–38. In *Water-Loss Investigations: Lake Mead Studies.* U.S. Geological Survey Professional Paper 298, Washington.

Harbeck, G. E., Kohler, M. A., Koberg, G. E., and others. 1958. *Water-Loss Investigations: Lake Mead Studies.* U.S. Geological Survey Professional Paper 298. Washington.

Hastings, J. R. 1960. *An Evaluation of Some Methods of Estimating Soil Moisture.* Unpublished paper. Dept. of Meteorology, The University of Arizona, Tucson.

Haugen, D. A., Barad, M. L., and Antanaitis, P. 1961. Values of Parameters Appearing in Sutton's Diffusion Models. *J. Meteorol.* 18:368–72.

Haurwitz, B. 1948. Insolation in Relation to Cloud Type. *J. Meteorol.* 5:110–13.

Haurwitz, B., and Austin, J. A. 1944. *Climatology.* McGraw-Hill Book Co., New York.

Hay, J. S., and Pasquill, F. 1959. Diffusion from a Continuous Source in Relation to the Spectrum and Scale of Turbulence, pp. 345–66. In *Advan. Geophys.* Vol. VI. Academic Press, New York.

Hellmann, G. 1919. Über die Bewegung der Luft in den untersten Schichten der Atmosphäre. *Proc. Akad. Wiss., Berlin,* pp. 404–16.

Heusser, C. J. 1961. Some Comparisons between Climatic Changes in Northwestern North America and Patagonia. *Ann. N.Y. Acad. Sci.* 95:642–57.

Hinzpeter, H. 1957. Die effektive Ausstrahlung und ihre Abhängigkeit von de Absorptionseigenschaften im Fenster der Wasserdamfbanden. *Z. Meteorol.* 11: 321–29.

Holland, J. Z. 1953. *A Meteorological Survey of the Oak Ridge Area.* Report ORO-99. Atomic Energy Commission, Washington.

Holzman, B. 1937. *Sources for Moisture for Precipitation in the United States.* U.S. Dept. Agri. Tech. Bull. 589. U.S. Government Printing Office, Washington.

Holzman, B. 1943. The Influence of Stability on Evaporation. *Ann. N.Y. Acad. Sci.* 44:13–18.

Hoover, M. D. 1944. Effect of Removal of Forest Vegetation upon Water Yields. *Trans. Am. Geophys. Union* 25:969–75.

Houghton, H. G. 1954. On the Annual Heat Balance of the Northern Hemisphere. *J. Meteorol.* 11:1–9.

House, G. J., Rider, N. E., and Tugwell, C. P. 1960. A Surface Energy-Balance Computer. *Quart. J. Roy. Meteorol. Soc.* 86:215–31.

Howell, L. G., and Martinez, J. D. 1957. Polar Movements as Indicated by Rock Magnetism. *Geophysics* 22:384–97.

Hubbs, C. L. 1957. Recent Climate History of California and Adjacent Areas, pp. 10–22. In *Recent Research in Climatology.* Committee on Research in Water Resources, University of California, La Jolla.

Hughes, G. H. 1963. *A Study of the Evaporation from Salton Sea, California.* Open-File Report. U.S. Geological Survey, Yuma.

Humphreys, W. J. 1940. *Physics of the Air.* McGraw-Hill Book Co., New York.

Hurley, P. M. 1959. *How Old Is the Earth?* Doubleday and Co., Inc., Garden City.

Hylckama, T. E. A., van. 1959. A Nomogram to Determine Monthly Potential Evapotranspiration. *Monthly Weather Rev.* 87:107–10.

Jacobs, W. C. 1951. Large-Scale Aspects of Energy Transformation over the Oceans, pp. 1057–70. In *Compendium of Meteorology.* American Meteorological Society, Boston.

Jardetsky, W. S. 1961. Investigations of Milankovitch and the Quaternary Curve of Effective Solar Radiation. *Ann. N.Y. Acad. Sci.* 95:418–23.

Johnson, F. S. 1954. The Solar Constant. *J. Meteorol.* 11:431–39.

Johnson, H. L., and Iriarte, B. 1959. *The Sun as a Variable Star.* Lowell Observatory Bull. No. 96. Lowell Observatory, Flagstaff.

Johnson, J. E. 1954. *Physical Meteorology.* John Wiley and Sons, Inc., New York.

Kaplan, L. D. 1960. The Influence of Carbon Dioxide Variations on the Atmospheric Heat Balance. *Tellus* 12:204–8.

Karlstrom, T. N. V. 1961. The Glacial History of Alaska: Its Bearing on Paleoclimatic Theory. *Ann. N.Y. Acad. Sci.* 95:290–340.

Kazanski, A. B., and Monin, A. S. 1956. Turbulentnost' v prizemnykh Inversiiakh. *Akad. Nauk SSSR, Izvestiia, Ser. Geofiz.* 1:79–86.

Kendrew, W. G. 1953. *The Climates of the Continents.* Clarendon Press, Oxford.

Kepner, R. A., Boelter, L. M. K., and Brooks, F. A. 1942. Nocturnal Wind Velocity, Eddy Stability and Eddy Diffusion above a Citrus Orchard. *Trans. Am. Geophys. Union* 23:239–49.

Kerr, R. S. 1960. *Water Resources Activities in the United States.* Committee Print No. 4, Select Committee on National Water Resources, U.S. Senate. U.S. Government Printing Office, Washington.

Kersten, M. S. 1949. *Thermal Properties of Soils.* Bull. 28, University of Minnesota Institute of Technology, Minneapolis.

Koberg, G. E. 1958. Energy-Budget Studies, pp. 20–29. In *Water-Loss Investigations: Lake Mead Studies.* U.S. Geological Survey Professional Paper 298, Washington.

Kohler, M. A. 1954. Lake and Pan Evaporation, pp. 127–48. In *Water-Loss Investigations: Lake Hefner Studies, Technical Report.* U.S. Geological Survey Professional Paper 269, Washington.

Kohler, M. A., Nordenson, T. J., and Baker, D. R. 1959. *Evaporation Maps for the United States.* U.S. Weather Bureau Technical Paper No. 37. Superintendent of Documents, Washington.

Kohler, M. A., Nordenson, T. J., and Fox, W. E. 1958. Pan and Lake Evaporation, pp. 38–60. In *Water-Loss Investigations: Lake Mead Studies.* U.S. Geological Survey Professional Paper 298, Washington.

Kozlowski, T. T. 1964. *Water Metabolism in Plants.* Harper and Row, New York.

Kraus, H. 1958. *Untersuchungen über den nächtlichen Energietransport und Energiehaushalt in der bodennahen Luftschicht bei der Bildung von Strahlungsnebeln.* Berichte Deutschen Wetterdienstes No. 48. Bad Kissingen.

Kuhn, P. M., and Suomi, V. E. 1958. Airborne Observations of Albedo with a Beam Reflector. *J. Meteorol.* 15:172–74.

Kuiper, G. P. 1957. Origin, Age, and Possible Ultimate Fate of the Earth, pp. 12–30. In *The Earth and Its Atmosphere.* Basic Books, Inc., New York.

Kuiper, G. P., and Middlehurst, B. M. 1961. *The Solar System*. Vol. III. *Planets and Satellites*. The University of Chicago Press, Chicago.

Kulp, J. L. 1961. Geologic Time Scale. *Science* 133:1105–14.

Kung, E. 1961. Derivation of Roughness Parameters from Wind Profile Data above Tall Vegetation, pp. 27–36. In *Studies of the Three-Dimensional Structure of the Planetary Boundary Layer*. Annual Report 1961. Dept. of Meteorology, University of Wisconsin, Madison.

Kung, E. C. 1963. Climatology of Aerodynamic Roughness Parameter and Energy Dissipation in the Planetary Boundary Layer of the Northern Hemisphere, pp. 37–96. In *Studies of the Effects of Variations in Boundary Conditions on the Atmospheric Boundary Layer*. Annual Report 1963. Dept. of Meteorology, University of Wisconsin, Madison.

Kung, E. C., Bryson, R. A., and Lenschow, D. H. 1964. Study of a Continental Surface Albedo on the Basis of Flight Measurements and Structure of the Earth's Surface Cover over North America. *Monthly Weather Rev.* 92:543–64.

Lachenbruch, A. H. 1957. Probe for Measurement of Thermal Conductivity of Frozen Soils in Place. *Trans. Am. Geophys. Union* 38:691–97.

Landsberg, H. 1958. *Physical Climatology*. Gray Printing Co., Inc., DuBois, Pa.

Lapple, C. E. 1964. Characteristics of Particles and Particle Dispersoids, pp. 622–23. In *Air Pollution*. Vol. I. Academic Press, New York.

Lauscher, F. 1934. Wärmeausstrahlung und Horizonteinengung. *Akad. Wissen., Vienna, Abt. IIa* 143:503–19.

Lee, C. H. 1942. Transpiration and Total Evaporation, pp. 259–330. In *Hydrology*. Dover Publications, Inc., New York.

Lee, R. L. 1963. *Evaluation of Solar Beam Irradiation as a Climatic Parameter of Mountain Watersheds*. Hydrology Paper No. 2. Colorado State University, Fort Collins.

Lee, W. H. K., and MacDonald, G. J. F. 1963. The Global Variation of Terrestrial Heat Flow. *J. Geophys. Res.* 68:6481–92.

Lehane, J. J., and Staple, W. J. 1953. Water Retention and Availability in Soils Related to Drought Resistance. *Can. J. Agr. Sci.* 33:265–73.

Lemon, E. 1962. *Energy and Water Balance in Plant Communities*. Interim Report No. 62-9. Agri. Research Service, Ithaca.

Lettau, H. H. 1951. Theory of Surface-Temperature and Heat-Transfer Oscillations near a Level Ground Surface. *Trans. Am. Geophys. Union* 32:189–200.

Lettau, H. H. 1952. The Present Position of Selected Turbulence Problems in the Atmospheric Boundary Layer, pp. 49–95. In *International Symposium on Atmospheric Turbulence in the Boundary Layer*. Geophysics Research Paper No. 19. AFCRC, Cambridge.

Lettau, H. H. 1954. Improved Models of Thermal Diffusion in the Soil. *Trans. Am. Geophys. Union* 35:121–32.

Lettau, H. H. 1962. Notes on Theoretical Models of Profile Structure in the Diabatic Surface Layer, pp. 195–226. In *Studies of the Three-Dimensional Structure of the Planetary Boundary Layer.* University of Wisconsin, Dept. of Meteorology, Madison.

Lettau, H. H., and Davidson, B. (eds.) 1957. *Exploring the Atmosphere's First Mile.* Vol. II. *Site Description and Data Tabulation.* Pergamon Press, Inc., New York.

Linke, F. 1931. Die nächtliche effektive Ausstrahlung unter verschiedenen Zenitdistanzen. *Meteorol. Z.* 48:25–31.

List, R. J. (ed.) 1958. *Smithsonian Meteorological Tables.* Smithsonian Institution, Washington.

Livingstone, B. E. 1910. Relation of Soil Moisture to Desert Vegetation. *Botan. Gaz.* 50:241–56.

Loennquist, O. 1954. Theoretical Verification of the Logarithmic Formula for the Relative Net Radiation to a Cloudless Sky. *Arkiv. Geofysik* 2:151–59.

London, J. 1957. *A Study of the Atmospheric Heat Balance.* Final Report, Project 131. New York University Dept. of Meteorology and Oceanography, New York.

London, J., and Haurwitz, M. W. 1963. Ozone and Sunspots. *J. Geophys. Res.* 68: 795–801.

Lowry, P. H. 1951. Microclimatic Factors in Smoke Pollution from Tall Stacks. *Meteorol. Monograph* 1 (4):24–29.

Lumley, J. L., and Panofsky, H. A. 1964. *The Structure of Atmospheric Turbulence.* John Wiley & Sons, Inc., New York.

Lund, I. A. 1964. *Indications of a Lunar Synodical Period in the Sunshine Observations for Boston, Massachusetts, and Columbia, Missouri.* Environmental Research Paper No. 9. AFCRL, Hanscom Field.

Lund, I. A. 1965. Indications of a Lunar Synodical Period in United States Observations of Sunshine. *J. Atmos. Sci.* 22:24–39.

McArdle, R. E. 1958. *Timber Resources for America's Future.* Forest Service Report No. 14. U.S. Government Printing Office, Washington.

McDonald, J. E. 1955. Paleoclimatology, pp. 196–200. In *Geochronology.* Physical Sciences Bull. No. 2. The University of Arizona, Tucson.

McDonald, J. E. 1961. On the Ratio of Evaporation to Precipitation. *Bull. Am. Meteorol. Soc.* 42:185–89.

McDonald, J. E. 1962. The Evaporation-Precipitation Fallacy. *Weather* 17:168–77, 216.

MacDonald, N. J., and Roberts, W. O. 1960. Further Evidence of a Solar Corpuscular Influence on Large-Scale Circulation at 300 mb. *J. Geophys. Res.* 65:529–34.

McIlroy, I. C., and Angus, D. E. 1964. Grass, Water, and Soil Evaporation at Aspendale. *Agr. Meteorol.* 1:201–24.

McVehil, G. E. 1964. Wind and Temperature Profiles near the Ground in Stable Stratification. *Quart. J. Roy. Meteorol. Soc.* 90:136–46.

Malkus, J. S. 1962. Large-Scale Interactions, pp. 88–294. In *The Sea*. Vol. I. John Wiley & Sons, Inc., New York.

Marlatt, W. E., Havens, A. V., Willits, N. A., and Brill, G. D. 1961. A Comparison of Computed and Measured Soil Moisture under Snap Beans. *J. Geophys. Res.* 66:535–41.

Martin, P. S. 1958. Pleistocene Ecology and Biogeography of North America, pp. 375–420. In *Zoogeography*. American Association for the Advancement of Science, New York.

Martin, P. S. 1963. *The Last 10,000 Years*. The University of Arizona Press, Tucson.

Mather, J. R. 1954. Investigation of Thornthwaite's Evapotranspiration Formula and Procedure, pp. 379–84. In *Estimating Soil Tractionability from Climatic Data.* Publications in Climatology, Vol. 7, No. 3. Drexel Institute of Technology, Centerton.

Mather, J. R. 1964. *Average Climatic Water Balance Data of the Continents.* Part V. *Europe.* Publications in Climatology, Vol. 17, No. 1. Drexel Institute of Technology, Centerton.

Meinardus, W. 1934. Niederschlagsverteilung auf der Erde. *Meteorol. Z.* 51:345–50.

Menzel, D. H. 1959. *Our Sun*. Harvard University Press, Cambridge.

Milankovitch, M. 1941. *Kanon der Erdbestrahlung und seine Anwendung auf des Eiszeitproblem.* Royal Serbian Academy, Belgrade.

Mintz, Y. 1954. *The Mean Geostrophic Poleward Flux of Angular Momentum and of Sensible Heat in the Winter and Summer of 1949.* Scientific Report No. 7. Dept. of Meteorology, University of California, Los Angeles.

Mitchell, J. M., Jr. 1961. Recent Secular Changes of Global Temperature. *Ann. N.Y. Acad. Sci.* 95:235–50.

Möller, F. 1951. Long-Wave Radiation, pp. 34–49. In *Compendium of Meteorology.* American Meteorological Society, Boston.

Molga, M. 1962. *Agricultural Meteorology*. Pt. II. *Outline of Agrometeorological Problems*. Office of Technical Services, Washington. (Transl. from Polish.)

Monin, A. S., and Budyko, M. I. 1964. About Heat-Balance Climatology. *Sov. Geograph.* 5:3–31.

Monin, A. S. and Obukhov, A. M. 1954. Dimensionless Characteristics of Turbulence in the Surface Layer. *Akad. Nauk SSSR. Geofis. Inst. Trudy* 151:163–87.

Moore, R. E. 1939. Water Conduction from Shallow Water Tables. *Hilgardia* 12: 383–426.

Munk, W. H., and Markowitz, W. 1960. North Pole Drifts 6 in. Each Year. In *New York Times*, Aug. 1.

Öpik, E. J. 1950. *Secular Changes of Stellar Structure and the Ice Ages.* Contribution No. 5. Armagh Observatory, N. Ireland.

Öpik, E. J. 1953. *A Climatological and Astronomical Interpretation of the Ice Ages and of the Past Variations of Terrestrial Climate.* Contribution No. 9. Armagh Observatory, N. Ireland.

Öpik, E. J. 1957. Ice Ages, pp. 152–73. In *The Earth and Its Atmosphere.* Basic Books, Inc., New York.

Öpik, E. J. 1958*a*. Climate and the Changing Sun. *Sci. Am.* 198 (6):85–92.

Öpik, E. J. 1958*b*. Solar Variability and Paleoclimatic Changes. *Irish Astron. J.* 5:97–109.

Ohman, H. L., and Pratt, R. L. 1956. *The Daytime Influence of Irrigation upon Desert Humidities.* Report EP-35, Environmental Protection Research Division. Quartermaster Research and Development Center, Natick.

Ol'dekop, E. M. 1911. On Evaporation from the Surface of River Basins. *Tr. I͡Ur'ev Obs.* Leningrad.

Opdyke, N. D. 1962. Paleoclimatology and Continental Drift, pp. 41–65. In *Continental Drift.* Academic Press, New York.

Palmer, W. C. 1965. *Meteorological Drought.* U.S. Weather Bureau Research Paper No. 45. Superintendent of Documents, Washington.

Palmer, W. C., and Havens, A. V. 1958. A Graphical Technique for Determining Evapotranspiration by the Thornthwaite Technique. *Monthly Weather Rev.* 86: 123–28.

Panofsky, H. A. 1956. Theories of Climatic Change. *Weatherwise* 9:183–87, 204.

Panofsky, H. A. 1961. An Alternative Derivation of the Diabatic Wind Profile. *Quart. J. Roy. Meteorol. Soc.* 87:109–10.

Panofsky, H. A. 1963. Determination of Stress from Wind and Temperature Measurements. *Quart. J. Roy. Meteorol. Soc.* 89:85–94.

Panofsky, H. A., and Townsend, A. A. 1964. Change of Terrain Roughness and the Wind Profile. *Quart. J. Roy. Meteorol. Soc.* 90:147–55.

Panofsky, H. A , Blackadar, A. K., and McVehil, G. E. 1960. The Diabatic Wind Profile. *Quart. J. Roy. Meteorol. Soc.* 86:390–98.

Pasquill, F. 1961. The Estimation of the Dispersion of Windborne Material. *Meteorol. Mag.* 90:33–49.

Pasquill, F. 1962. *Atmospheric Diffusion.* D. Van Nostrand Company Ltd., New York.

Pauly, K. 1952. The Cause of the Great Ice Ages. *Sci. Monthly* 75(8):89–98.

Pèivé, T. L. 1960. Glacial History of the McMurdo Sound Region, Antarctica. *IGY Bull.* 41(2):1–7.

Peixoto, J. P. 1960. *Hemispheric Temperature Conditions during the Year 1950.* Scientific Report No. 4, Dept. of Meteorology, Massachusetts Institute of Technology, Cambridge.

Pelton, W. L., King, K. M., and Tanner, C. B. 1960. An Evaluation of the Thornthwaite and Mean Temperature Methods for Determining Potential Evapotranspiration. *Agron. J.* 52:387–95.

Penman, H. L. 1948. Natural Evaporation from Open Water, Bare Soil, and Grass. *Proc. Roy. Soc. London* A193:120–45.

Penman, H. L. 1956*a*. Estimating Evaporation. *Trans. Am. Geophys. Union* 37:43–50.

Penman, H. L. 1956*b*. Evaporation: An Introductory Survey. *Neth. J. Agr. Sci.* 1:9–29, 87–97, 151–53.

Penman, H. L. 1961. Weather, Plant, and Soil Factors in Hydrology. *Weather* 16:207–19.

Penman, H. L., and Long, I. F. 1960. Weather in Wheat: An Essay in Micro-Meteorology. *Quart. J. Roy. Meteorol. Soc.* 86:16–50.

Petterssen, S. 1959. *On the Influence of Heat Exchanges on Motion and Weather Systems.* Scientific Report No. 10, Contract AF-19(604)-2179, University of Chicago, Chicago.

Pettersson, O. 1914. *Climatic Variations in Historic and Prehistoric Time.* Svenska Hydrogr. Biol. Komm., Skriften, Vol. V.

Plass, G. N. 1956. Effect of Carbon Dioxide Variations on Climate. *Am. J. Phys.* 24:376–87.

Plass, G. N. 1957. The Carbon Dioxide Theory of Climatic Change, pp. 81–92. In *Recent Research in Climatology.* Committee on Research in Water Resources, University of California, La Jolla.

Plass, G. N. 1961. The Influence of Infrared Absorptive Molecules on the Climate. *Ann. N.Y. Acad. Sci.* 95:61–71.

Priestley, C. H. B. 1955. Free and Forced Convection in the Atmosphere near the Ground. *Quart. J. Roy. Meteorol. Soc.* 81:139–43.

Priestley, C. H. B. 1959. *Turbulent Transfer in the Lower Atmosphere.* University of Chicago Press, Chicago.

Pruitt, W. O., and Angus, D. E. 1960. Large Weighing Lysimeters for Measuring Evapotranspiration. *Trans. Am. Soc. Agr. Eng.* 3:13–18.

Raethjen, P. 1950. Kurzer Abriss der Meteorologie dynamisch gesehen. Teil II. Wärmehaushalt der Atmosphäre. *Geophysik. Einzel.*, pp. 103–52.

Reitan, C. H. 1960*a*. Mean Monthly Values of Precipitable Water over the United States. *Monthly Weather Rev.* 88:25–35.

Reitan, C. H. 1960*b*. Distribution of Precipitable Water Vapor over the Continental United States. *Bull. Am. Meteorol. Soc.* 41:79–87.

Rense, W. A. 1961. Solar Radiation in the Extreme Ultraviolet Region of the Spectrum and Its Effect on the Earth's Upper Atmosphere. *Ann. N.Y. Acad. Sci.* 95:33–38.

Reynolds, O. 1883. An Experimental Investigation of the Circumstances Which Determine Whether the Motion of Water Shall Be Direct or Sinuous, and of the Law of Resistance in Parallel Channels. *Phil. Trans. Roy. Soc.* 174:935–82.

Richards, L. A., and Richards, S. J. 1957. Soil Moisture, pp. 49–60. In *Soil.* Superintendent of Documents, Washington.

Rider, N. E. 1954. Eddy Diffusion of Momentum, Water Vapour, and Heat near the Ground. *Phil. Trans. Roy. Soc. London* A246:481–501.

Rider, N. E., and Robinson, G. D. 1951. A Study of the Transfer of Heat and Water Vapour Above a Surface of Short Grass. *Quart. J. Roy. Meteorol. Soc.* 77:375–401.

Rider, N. E., Philip, J. R., and Bradley, E. F. 1963. The Horizontal Transport of Heat and Moisture—a Micrometeorological Study. *Quart. J. Roy. Meteorol. Soc.* 89:507–31.

Roach, W. T. 1961. The Absorption of Solar Radiation by Water Vapour and Carbon Dioxide in a Cloudless Atmosphere. *Quart. J. Roy. Meteorol. Soc.* 87:364–73.

Roberts, O. F. T. 1923. The Theoretical Scattering of Smoke in a Turbulent Atmosphere. *Proc. Roy. Soc. London, Ser. A*, 104:640–54.

Robinson, G. D. 1950. Notes on the Measurement and Estimation of Atmospheric Radiation—2. *Quart. J. Roy. Meteorol. Soc.* 76:37–51.

Robitzsch, M. 1926. Strahlungsstudien Ergebnisse. *Lindenberg, Preussisches Aero. Obs., Arbeiten* 15:194.

Rombakis, S. 1947. Über die Verbreitung von Pflanzensamen und Sporen durch turbulente Luftströmungen. *Z. Meteorol.* 1:359–63.

Rossby, C. G. 1945. The Scientific Basis of Modern Meteorology, pp. 502–29. In *Handbook of Meteorology*. McGraw-Hill Book Co., Inc., New York.

Rossby, C. G., and Montgomery, R. B. 1935. *The Layer of Frictional Influence in Wind and Ocean Currents*. Papers in Physical Oceanography and Meteorology, M.I.T. and Woods Hole Oceanographic Institution, Vol. 3, No. 3.

Rubin, M. J. 1964. Antarctic Weather and Climate, pp. 461–78. In *Research in Geophysics*. Vol. II. *Solid Earth and Interface Phenomena*. M.I.T. Press, Cambridge.

Rubin, M. J., and Weyant, W. S. 1963. The Mass and Heat Budget of the Antarctic Atmosphere. *Monthly Weather Rev.* 91:487–93.

Runcorn, S. K. 1956. Paleomagnetic Survey in Arizona and Utah. *Bull. Geol. Soc. Am.* 67:301–16.

Runcorn, S. K. 1962. Paleomagnetic Evidence for Continental Drift and Its Geophysical Cause, pp. 1–40. In *Continental Drift*. Academic Press, New York.

Russell, M. B. 1939. Soil Moisture Sorption Curves for Four Iowa Soils. *Soils Sci. Soc. Amer. Proc.* 4:51–54.

Sanderson, I. T. 1960. Riddle of the Frozen Giants, pp. 39, 82–83. In *Saturday Evening Post*, Jan. 16.

Sauberer, F. 1951. Das Licht im Boden. *Wetter und Leben* 3:40–44.

Schmidt, W. 1908. Absorption der Sonnenstrahlung im Wasser. *Sitzber. Akad. Wiss., Wien (2A)* 117:237–53.

Schmidt, W. 1918. Wirkungen des Luftaustauches auf das Klima und den täglichen Gang der Lufttemperatur in der Höhe. *Sitzber. Akad. Wiss. Wien* 127:1942–57.

Schove, D. J. 1961. Solar Cycles and the Spectrum of Time since 200 B.C. *Ann. N.Y. Acad. Sci.* 95:107–23.

Schreiber, P. 1904. Über die Beziehungen zwischen dem Niederschlag und der Wasserführung der Flüsse in Mitteleuropa. *Meteorol. Z.* 21:441–52.

Schwarzbach, M. 1963. *Climates of the Past, An Introduction to Paleoclimatology.* D. Van Nostrand Co., Ltd., New York.

Schwarzschild, M. 1958. *Structure and Evolution of the Stars.* Princeton University Press, Princeton.

Seale, R. L., and Couchman, J. C. 1961. *Empirical Correlations of Atmospheric Dispersal Data.* FZM-2278. Convair, Fort Worth.

Sears, P. B. 1958. Environment and Culture in Retrospect, pp. 77-84. In *Climate and Man in the Southwest.* The University of Arizona Press, Tucson.

Sellers, W. D. 1960. Precipitation Trends in Arizona and Western New Mexico, pp. 81–94. In *Proceedings of the 28th Annual Western Snow Conference.* U.S. Soil Conservation Service, Fort Collins.

Sellers, W. D. 1962. A Simplified Derivation of the Diabatic Wind Profile. *J. Atmos. Sci.* 19:180–81.

Sellers, W. D. 1964*a*. The Energy Balance of the Atmosphere and Climatic Change. *J. Appl. Meteorol.* 3:337–39.

Sellers, W. D. 1964*b*. Potential Evapotranspiration in Arid Regions. *J. Appl. Meteorol.* 3:98–104.

Sellers, W. D., and Hodges, C. N. 1962. The Energy Balance of Non-Uniform Soil Surfaces. *J. Atmos. Sci.* 19:482-91.

Shapiro, R. 1958. Some Observations of the Persistence of the Surface Pressure Distribution. *J. Meteorol.* 15:435–39.

Shapiro, R. 1959. A Comparison of the Response of the North American and European Surface Pressure Distributions to Large Geomagnetic Disturbances. *J. Meteorol.* 16:569–72.

Shaw, R. H., and McComb, A. L. 1959. A Comparison of the Gunn-Bellani Radiation Integrator and the Eppley Pyrheliometer. *Forest Sci.* 5:234–36.

Simpson, G. C. 1929. The Distribution of Terrestrial Radiation. *Mem. Roy. Meteorol. Soc.* 3:53–78.

Simpson, G. C. 1934. World Climate during the Quaternary Period. *Quart. J. Roy. Meteorol. Soc.* 60:425–78.

Simpson, G. C. 1957. Further Studies in World Climate. *Quart. J. Roy. Meteorol. Soc.* 83:459–81.

Simpson, G. C. 1959. World Temperatures during the Pleistocene. *Quart. J. Roy. Meteorol. Soc.* 85:332–49.

Sinclair, J. G. 1922. Temperatures of the Soil and Air in a Desert. *Monthly Weather Rev.* 50:142–44.

Slatyer, R. O., and McIlroy, I. C. 1961. *Practical Microclimatology.* CSIRO, Melbourne.

Special Committee for the International Geophysical Year. 1958. Radiation Instruments and Measurements. *Ann. Intern. Geophys. Yr.* 5:367–466.

Starr, V. P., and Peixoto, J. P. 1958. On the Global Balance of Water Vapor and the Hydrology of Deserts. *Tellus* 10:188–94.

Starr, V. P., and Peixoto, J. P. 1964. The Hemispheric Eddy Flux of Water Vapor and Its Implications for the Mechanics of the General Circulation. *Arch. Meteorol. Geophys. Bioklimatol. Ser. A.* 14:111–30.

Starr, V. P., and White, R. M. 1955. Direct Measurement of the Hemispheric Poleward Flux of Water Vapor. *J. Marine Res.* 14:217–25.

Starr, V. P., Peixoto, J. P., and Livadas, G. C. 1958. On the Meridional Flux of Water Vapor in the Northern Hemisphere. *Geofis. Pura Appl.* 39:174–85.

Stepanova, N. 1963. The World's Lowest Temperature Record. *Weatherwise* 16:268–69.

Stoval, J. W., and Brown, H. E. 1955. *The Principles of Historical Geology*. Ginn and Company, Boston.

Strom, G. H. 1962. Atmospheric Dispersion of Stack Effluents, pp. 118–95. In *Air Pollution*. Vol. I. Academic Press, New York.

Süssenberger, E. 1935. Die nächtliche effektive Ausstrahlung unter verschiedenen Zenitdistanzen. *Meteorol. Z.* 52:129–32.

Suomi, V. E., Staley, D. O., and Kuhn, P. M. 1960. A Direct Measurement of Infra-Red Radiation Divergence to 160 mb. *Quart. J. Roy. Meteorol. Soc.* 84:134–41, 472–74.

Sutton, O. G. 1953. *Micrometeorology*. McGraw-Hill Book Co., New York.

Sverdrup, H. U. 1936. *The Eddy Conductivity of the Air over a Smooth Snow Field*. Vol. 11, No. 7. Geofysiske Publikasjoner, Oslo.

Swan, J. B., Federer, C. A., and Tanner, C. B. 1961. *Economical Radiometer Performance, Construction, and Theory*. Soils Bull. 4. University of Wisconsin, Madison.

Swinbank, W. C. 1955. *An Experimental Study of Eddy Transports in the Lower Atmosphere*. Technical Paper No. 2. C.S.I.R.O. Div. Meteorological Physics, Melbourne.

Swinbank, W. C. 1963. Long-wave Radiation from Clear Skies. *Quart. J. Roy. Meteorol. Soc.* 89:339–48. (Also, 90:488–93.)

Swinbank, W. C. 1964. The Exponential Wind Profile. *Quart. J. Roy. Meteorol. Soc.* 90:119–35.

Tank, W. 1957. The Use of Large-Scale Parameters in Small-Scale Diffusion Studies. *Bull. Am. Meteorol. Soc.* 38:6–12.

Tanner, C. B. 1963. *Basic Instrumentation and Measurements for Plant Environment and Micrometeorology*. Soils Bull. 6. University of Wisconsin, Madison.

Tanner, C. B., and Pelton, W. L. 1960*a*. Potential Evapotranspiration Estimates by the Approximate Energy Balance Method of Penman. *J. Geophys. Res.* 65: 3391-3413.

Tanner, C. B., and Pelton, W. L. 1960*b*. *Energy Balance Data—Hancock, Wisconsin.* Soils Bull. 2. University of Wisconsin, Madison.

Tanner, C. B., Businger, J. A., and Kuhn, P. M. 1960. The Economical Net Radiometer. *J. Geophys. Res.* 65:3657–67.

Taylor, S. A. 1957. Use of Moisture by Plants, pp. 61-66. In *Soils*. Superintendent of Documents, Washington.

Terasmae, J. 1961. Notes on Late-Quaternary Climatic Changes in Canada. *Ann. N.Y. Acad. Sci.* 95:658–75.

Thornthwaite, C. W. 1948. An Approach toward a Rational Classification of Climate. *Geogr. Rev.* 38:55–94.

Thornthwaite, C. W., and Mather, J. R. 1955. *The Water Balance.* Publications in Climatology, Vol. 8, No. 1. Drexel Institute of Technology, Centerton.

Thornthwaite, C. W., and Mather, J. R. 1957. *Instructions and Tables for Computing Potential Evapotranspiration and the Water Balance.* Publications in Climatology, Vol. 10, No. 3. Drexel Institute of Technology, Centerton.

Thornthwaite, C. W., and Mather, J. R. 1962. *Average Climatic Water Balance Data of the Continents.* Pt. I. *Africa.* Publications in Climatology, Vol. 15, No. 2, Drexel Institute of Technology, Centerton.

Trewartha, G. T. 1954. *An Introduction to Climate.* McGraw-Hill Book Co., Inc., New York.

Trewartha, G. T. 1961. *The Earth's Problem Climates.* University of Wisconsin Press, Madison.

Uhlig, S. 1954. *Berechnung der Verdunstung aus klimatologischen Daten.* Mitteilungen des Deutschen Wetterdienstes, Vol. 1, No. 6. Bad Kissingen.

U.S. Weather Bureau, 1957. *Upper-Air Climatology of the United States.* U.S. Weather Bureau Technical Paper No. 32. U.S. Government Printing Office, Washington.

U.S. Weather Bureau 1959. *Climatography of the United States,* No. 60. U.S. Government Printing Office, Washington.

Urey, H. C. 1952. *The Planets.* Oxford University Press, London.

Urey, H. C., Lowenstam, H. A., Epstein, S., and McKinney, C. R. 1951. Measurements of Paleotemperatures and Temperatures of the Upper Cretaceous of England, Denmark, and the Southeastern United States. *Bull. Geol. Soc. Am.* 62:399–416.

Vehrencamp, J. E. 1951. *An Experimental Investigation of Heat and Momentum Transfer at a Smooth Air-Earth Interface.* Dept. of Engineering, University of California, Los Angeles.

Vehrencamp, J. E. 1953. Experimental Investigation of Heat Transfer at an Air-Earth Interface. *Trans. Am. Geophys. Union* 34:22–30.

Vening Meinesz, F. A. 1962. Thermal Convection in the Earth's Mantle, pp. 145–76. In *Continental Drift.* Academic Press, New York.

Vitkevich, V. I. 1963. *Agricultural Meteorology*. Office of Technical Service, Washington. Translation prepared within the Special Foreign Currency Science Information (SFCSI) Program of the National Science Foundation.

Vries, D. A. de. 1958. The Thermal Behaviour of Soils, pp. 109–13. In *Climatology and Microclimatology*. UNESCO, Paris.

Vries, D. A. de. 1963. Thermal Properties of Soils, pp. 210–35. In *Physics of Plant Environment*. North-Holland Publishing Co., Amsterdam.

Vries, D. A. de, and Peck, A. J. 1957. On the Cylindrical Probe Method of Measuring Thermal Conductivity with Special Reference to Soils. *Austr. J. Phys.* 11:255–71.

Waldmeier, M. 1956–1964. Geomagnetic and Solar Data. *J. Geophys. Res.* Vol. 61–70.

Wark, D. Q., Yamamoto, G., and Lienesch, J. H. 1962. Methods of Estimating Infrared Flux and Surface Temperature from Meteorological Satellites. *J. Atmos. Sci.* 19:369–84.

Webb, E. K. 1960. *Evaporation from Lake Eucumbene*. Technical Paper No. 10. C.S.I.R.O. Div. Meteorological Physics, Melbourne.

Wexler, H. 1955. *Meteorology and Atomic Energy*. U.S. Government Printing Office, Washington.

Wexler, H. 1960. Possible Causes of Climatic Fluctuations, pp. 93–95. In *Dynamics of Climate*. Pergamon Press, New York.

Wijk, W. R. van. 1963. General Temperature Variations in a Homogeneous Soil, pp. 144–70. In *Physics of Plant Environment*. North-Holland Publishing Co., Amsterdam.

Wijk, W. R. van, and Derksen, W. J. 1963. Sinusoidal Temperature Variation in a Layered Soil, pp. 171–209. In *Physics of Plant Environment*. North-Holland Publishing Co., Amsterdam.

Wijk, W. R. van, and Vries, D. A. de. 1963. Periodic Temperature Variations in a Homogeneous Soil, pp. 102–43. In *Physics of Plant Environment*. North-Holland Publishing Co., Amsterdam.

Willett, H. C. 1953. Atmospheric and Oceanic Circulation as Factors in Glacial-Interglacial Changes of Climate, pp. 51–71. In *Climatic Change*. Harvard University Press, Cambridge.

Willett, H. C. 1961. The Pattern of Solar Climatic Relationships. *Ann. N.Y. Acad. Sci.* 95:89–106.

Willett, H. C. 1962. The Relationship of Total Atmospheric Ozone to the Sunspot Cycle. *J. Geophys. Res.* 67:661–70.

Willett, H. C. 1964. Evidence of Solar-Climatic Relationships, pp. 123–51. In *Weather and Our Food Supply*. Iowa State University, Ames.

Willett, H. C. 1965. Solar-Climatic Relationships in the Light of Standardized Climatic Data. *J. Atmos. Sci.* 22:120–36.

Woerkom, A. J. J. van. 1953. The Astronomical Theory of Climatic Changes, pp. 147–57. In *Climatic Change.* Harvard University Press, Cambridge.

Wolbach, J. 1955. The Insufficiency of Geographical Causes of Climatic Change, pp. 107–16. In *Climatic Change.* Harvard University Press, Cambridge.

World Meteorological Organization. 1963. *Guide to Meteorological Instrument and Observing Practices.* 2d ed., Suppl. 1. World Meteorol. Organ., Geneva.

Wright, H. E., Jr. 1963. Retrospect on the Contributions of Geological Studies to Problems of the Climate of the Eleventh and Sixteenth Centuries, pp. 31–36. In *Proceedings of the Conference on the Climate of the Eleventh and Sixteenth Centuries.* NCAR Technical Note 63-1, Natl. Center for Atmos. Research, Boulder.

Wright, J. L., and Lemon, E. R. 1962. *Estimation of Turbulent Exchange within a Corn Crop Canopy at Ellis Hollow.* Interim Report 62-7. N.Y.S. College of Agri., Ithaca.

Wüst, G. 1954. Gesetzmässige Wechselbeziehungen zwischen Ozean und Atmosphäre in der zonalen Verteilung von Oberflächensalzgehalt, Verdunstung, und Niederschlag. *Arch. Meteorol., Geophys., Bioklimatol. Ser. A.* 7:305–28.

Yamamoto, G. 1952. On a Radiation Chart. Tôhoku Univ., Science Rep., Series 5, *Geophysics* 4:9–23.

Yamamoto, G. 1959. Theory of Turbulent Transfer in Non-Neutral Conditions. *J. Meteorol. Soc. Japan.* 37:60–70.

Yamamoto, G., and Kondo, J. 1964. Evaporation from Lake Towada. *J. Meteorol. Soc. Japan.* 42:85–96.

Yamamoto, G., and Shimanuki, A. 1964. Profiles of Wind and Temperature in the Lowest 250 Meters in Tokyo. Tôhoku Univ., Science Rep., Series 5, *Geophysics* 15:111–14.

Yocum, C. S., Allen, L. H., and Lemon, E. R. 1962. *Solar Radiation Balance and Photosynthetic Efficiency.* Interim Report No. 62-3. Agri. Research Service, Ithaca.

Zeuner, F. E. 1959. *The Pleistocene Period.* Hutchinson Science and Technology, London.

Zeuner, F. E. 1961. The Sequence of Terraces of the Lower Thames and the Radiation Chronology. *Ann. N.Y. Acad. Sci.* 95:377–80.

AUTHOR INDEX

SUBJECT INDEX